A New Revolution in Physics

Why a New Revolution in Physics is Essential and why it is Inevitable

(Book 1 of 10)

I. P. K. R. Hirwani

Copyright © 2019 I.P.K.R. Hirwani

All rights reserved.

ISBN: 9781080041367

DEDICATION

This book is dedicated to my father,
Shri. K. S. Hirwani and my mother Smt. R. K. Hirwani.

CONTENTS

	Acknowledgments	ix
	Important quotes	xi
1	What is the aim of these Books	1
2	Why so many Books?	2
3	About this Book	4
4	Who should read these Books	5
5	Why one should read these Books?	6
6	About the Author	9
	6.1 Who is the Author?	9
	6.2 What is the motive of the Author	9
	6.3 Is the Author qualified for the subjects?	11
	6.4 Why should anybody care for a non-mainstream author's views and claims?	19
	6.5 Are the main ideas / theories original s	20
	6.6 The source of Author's ideas and theories	20
7	Are these Books based on scientific principles?	21
8	What are scientific principles?	22
9	What are the main claims of the Author?	23

10	Why shouldn't the claims of the author be rejected?	24
11	What is "A New Revolution in Physics"?	28
12	Why the New revolution in Physics is essential?	32
13	What is the accepted main problem of the fundamental theories of Modern Physics?	40
14	Why the new revolution in Physics is inevitable?	50
15	Very brief history of development of the fundamental theories of Modern Physics with present accepted views	53
	15.1 Special Theory of Relativity	54
	15.2 General Theory of Relativity	57
	15.3 Quantum Mechanics	61
16	Why different fundamental theories of the Modern Physics were required to be created?	69
17	What these fundamental theories of Modern Physics claim to explain?	71
18	Are the claims of the fundamental theories of Modern Physics scientifically correct?	74
19	Are the present accepted fundamental theories developed on scientific principles?	77
20	The Classical Physics concepts which the Modern Physics has changed	78
21	The new concepts which the Modern Physics introduced	88

22	The concepts or problems which the fundamental theories of Modern Physics do not explain but the new Quassical Physics explains	96
23	What about the experimental confirmations of the fundamental theories?	102
24	What about the technological 'proofs' of Relativity and Quantum Physics - GPS, Lasers, LEDs, Cell phones and all sorts of electronics etc.?	103
25	The accepted concepts of the fundamental theories of Modern Physics which are non-scientific	106
26	What is the basis of the Quassical Physics? How its basis is different from the basis of the presently accepted physics	108
27	How the Quassical Physics affects the Particle physics	111
28	How the Quassical Physics affects the mathematics	112
29	How the Quassical Physics affects the Physics Establishment, the mainstream physicists and the peer review process?	113
30	The Next book to be published "The Collapse of the Quantum Physics"	115
31	CONCLUSIONS	116

ACKNOWLEDGMENTS

I am grateful to my parents and family for supporting me in every respect.
I am grateful to my family, wife Seema, sons Kshitij and Abhyuday for supporting me in my work and bearing the pains I caused them in the course of this work.

I am grateful to Mr. Anand Doctor for his important crucial support for this work.

IMPORTANT QUOTES

"In the science the authority of thousands of opinions is not worth as much as one tiny spark of reason in an individual man."
-Galileo Galilei

"Truth is found if ever in the simplicity, never in the multiplicity and confusion of things"
-Sir Isaac Newton

"Convictions are more dangerous enemies of the truth than lies."
-Friedrich Nietzsche

"Beware of false knowledge, it is more dangerous than ignorance."
-George Bernard Shaw

1
WHAT IS THE AIM OF THESE BOOKS

The aim of these books is to bring the crises in the Modern Physics to the forefront, to scientifically prove that the crises exist and has become uncontrollable, to pinpoint the cause of the crises and to provide the scientific solution of the crises, which requires review of the foundation and the fundamental concepts of the whole physics, at least from the beginning of the 20th century.

The aim of these books is also to highlight the non-scientific culture being followed in the Physics Establishment, I call it "The System", which were forced by the founders of the fundamental theories, it's time to review it and to follow the scientific culture.

The aim is to provide environments to the mainstream physicists for introspection, to ask questions what are scientific principles, are they following the scientific principles for the progress of the Physics? How much the Modern Physics has deviated from its intended purpose? How they can contribute to restore the true sense of Physics in Physics.

The aim is to scientifically pin point what exactly has gone wrong in the development of the fundamental theories of the Modern Physics, in philosophical, conceptual, theoretical, mathematical and interpretational fronts, what are their effects and how to correct them.

The aim is to provide a new solution to majority of the problems of the Modern Physics as well some of the problems of the Classical Physics in the form of new theory of Physics completely based on the empirical findings and true scientific and philosophical principles.

Finally, the aim is to respond to the present crises in Physics and initiate a new scientific revolution in Physics to restore its heart and true characters.

2
WHY SO MANY BOOKS?

There are lots of problems in almost every field of Modern Physics, each book highlights and deals with one field of the Modern Physics in details. These books covers in depth analysis of non-scientific steps in all fields of Modern Physics and bring them in the forefront and makes them so blatant that it becomes hard to neglect thereafter. One book is necessary to concentrate on one particular field.

Some people try to raise their voice against some fundamental theories to the Physics Establishment or The System, but they do not have enough hard core scientific evidences or theoretical arguments which can make a scratch on the System, and even some people have presented some scientific evidences which contradicts the accepted fundamental theories, but the System has become so powerful that some isolated one or two or a few hard scientific evidences fails to make any mark. To make a mark, lots of evidences and lots of hard scientific arguments are required which is difficult to contain in a single book, it's better to segregate in a series of books with a single theme.

A single large book in this regard will be thrown straight into a dustbin, without even opening it, nobody wants the equilibrium to be disturbed and become uncomfortable, but for the sake of physics, highlighting the crises has become extremely essential, as we have been through the crises for over a century. This is the reason for multiple books, so that only one issue can be focused deeply and critically analyzed at a time.

Another reason for many books is that, the author is coming in this field from nowhere, and making impossible tall claims, let us judge scientifically the author's work from the initial books, if it worth, then judge the main work – "The Concepts, The Dawn of the New Physics - The Quassical Physics", which will initiate a new Revolution in Physics, which has been due for over a century and has now become essential and inevitable.

3
ABOUT THIS BOOK

This book is the first book of the series "A New Revolution in Physics" which contains 10 books. "Why A New Revolution in Physics is Essential and why it is Inevitable?" As the title of the first book indicates, in this book a broad view is given about the present state of physics, some critical views are presented about the present accepted understandings and fundamental theories; it seems scientific methods are not followed in the development of present accepted fundamental theories, personal views dominated over scientific reasoning. I believe this led the physics into a non-scientific path. This book mainly highlights the reasons why the new revolution has now become essential and inevitable. There are several critical examples of such bias which changed the direction of the development of physics, in this book, the first book of the series, some lists of important and critical concepts and experiments are given which decided the direction of progress, every items of the lists are discussed in details in following other books which unequivocally shows a clear non-scientific path taken by the Modern Physics responsible for the crises.

4
WHO SHOULD READ THESE BOOKS

In my view, everyone who is even a little bit interested in how the Nature or the real physical world works should read these books.

Everyone who is deeply interested and / or deeply connected with Physics, particularly, who are responsible for guiding and maintaining the progress of physics.

Everyone who felt the initial discomfort when first encountered the theory of Relativity and the theory of Quantum Mechanics, which is still alive within her / him in some corner and wants the cure for the discomfort.

Everyone who is troubled by the incompatibility of the fundamental theories of Modern Physics, i.e., the incompatibility of General theory of Relativity and the theory of Quantum Mechanics. Everyone who wants to know what the reasons for the incompatibility are and how the new physics, the Quassical Physics, solves the problem.

Everyone who is troubled by the mysteries and "magic" of modern physics like the metric tensor, the curvature of space-time, the twin paradox, the wave function, the Schrodinger's cat, the virtual particles, the double slit experiment, the quantum jump, spin etc.

Everyone who is troubled by the vast zoo of fundamental particles and virtual particles.

Finally, these books are must for mainstream physicists, because it deals with their subject with a scientific but contrarian views and intended to bring a new revolution in their fields.

.

5
WHY ONE SHOULD READ THESE BOOKS?

The first reason to read these books is to get an unbiased and critical view of the present fundamental theories of physics from someone who is not part of the mainstream Physics Establishment, not being the part of mainstream Physics Establishment has the advantage of becoming dead critical about the fundamental theories of physics. Being dead critical is a scientific requirement for scientific progress of physics and related theories which has been compromised for over a century, that's why true scientific views about the fundamental theories never emerged from the Establishment.

Another reason is to know how and why in the development of modern fundamental theories of physics the non-scientific steps crept in and lead to a wrong course, and to witness the course correction of a century old wrong path Modern Physics has adopted.

Another reason is to witness the New Revolution in Physics, to be a part of the revolution, witness the fall of insurmountable Relativity and Quantum Physics purely on scientific experimental grounds, and to witness the restoration of the heart of physics and the resurrection of the Classical Physics in its new slightly modified magnificent form, the new incarnation of the Grand Old Classical Physics, without Relativity and Quantum Physics – The Quassical Physics, which address the deep buried core issues of absolute reference frame, causality, energy, action at a distance, inertia, fields, mass, momentum, centrifugal force, heat etc. It also redefines the concepts of space, time, simultaneity, the cause of gravity, equivalence of inertial and gravitational mass, which are distorted by Relativity and explains scientifically the problems of the stability if atom, the atomic spectral lines, the quantum jump etc. which are explained non-scientifically by quantum mechanics. The Quassical Physics explains them without using any postulates and concepts of Relativity and Quantum Mechanics.

Another reason, a very important reason, is how the Physics is emancipated from the abstract mathematics and therefore, to get rid of the discomfort caused by the highly abstract mathematical theories of physical reality, the core of the new physics is intuitive and extremely simple and least mathematical. To get the physical understandings of many basic concepts of physics which are explained by the Modern Physics in highly mathematical way without providing any physical understanding, like what is gravity, how things move under gravity, what are orbitals, how electrons are distributed in the atom, how the interference pattern in a double slit experiment is explained, how a photon carries momentum, what is time etc.

Another reason is to know the solution of the biggest problem of the Modern Physics – the incompatibility problem, how the new physics provides a seamless single theory for the world of small objects to the world of large objects completely consistent with the world of intermediate objects i.e., our daily observable world.

Another reason is mainly for the mainstream physicists, to get unbiased, scientific and critical views against their own professional views, which is impossible to get from within the system. To know they have been living in the crises, to realize that the crises are much bigger than they could have imagined -Relativity and Quantum physics are founded on non-scientific and non-philosophical principles and that's why they are in big crises. The still bigger problem is that, majority of them are not even aware that they are in crises, to get out of the crises, and the most important personal matter for them, to stop wasting their effort and life on wrong theories. They may achieve professional success but, in my view, they are aware of the hollowness of the whole business on personal level, they must read these books to remove this hollowness form their personal life and contribute to remove the hollowness from the physics also.

Another reason is to witness one's first instinct was correct, when encountered the theory of relativity and quantum physics. In general every student of science when first encounters the presently accepted fundamental theories of Physics - the theory of Relativity and the theory of Quantum Physics, becomes shocked because of the difficulty to co-relate the core nature of the theories and the nature of the world of physical reality of his own experience, the mismatch is very obvious, initially this is very uncomfortable, everyone passes through this experience and the period of discomfort, but as time passes it becomes very difficult to remain un influenced by the copious amount of information lurking through every possible channels in support and veneration of these theories as the greatest achievements of human mind. Gradually the intensity of the initial discomfort dies down and the majority of them accept the projected reality on the belief that the esteemed scientific journals and the results of the mega experiments conducted by the elite laureates are beyond doubts. Everybody do not have the resources to dig deep to check by their own. People get converted, and in such circumstances this is natural thus to become the part of the accepted Physics Establishment, or the System, unknowingly, but, I believe the initial discomfort is still alive in howsoever diminished intensity and would like to seek the answers in non-professional personal capacity. To get the answers one should read these books.

The most important feature of all the books in this series, except only this first book, where mainly a broad perspective is given, is that, all the arguments, reasoning, claims, proposals, predictions and conclusions are based on firm experimental basis, they follow strict scientific methods and are open for any and every possible questions and further scientific scrutiny on philosophical, theoretical, conceptual, mathematical and interpretational fronts. They are based on empirical evidences of real physical world obtained through the experiments which have already been performed.

6
ABOUT THE AUTHOR

It is not usual to find a full chapter dedicates to the author, but the circumstances and the context demands it. Although this is a chapter about author but it is more about the circumstances which made a normal person an author of such an important subject.

6.1   WHO IS THE AUTHOR?

I. P. K. R. Hirwani, I. & P. stands for the author's name, K stands for author's father's name Mr. K S Hirwani and R stands for author's mother's name Mrs. R. K. Hirwani. The author is just a little bit more curious to understand about the working of the Nature, he has been involved in fundamentals of Physics and Mathematics for about three decades and completely dedicated for last three years.

6.2   WHAT IS THE MOTIVE OF THE AUTHOR

I am interested in how the physical world works; these books are the result of my long solo self-study, for my own satisfaction. I have found many critical errors and non-scientific steps in the formation of fundamental theories of physics, to my surprise; I have discovered two great epicycles in the Modern Physics - the theory of Relativity and the theory of Quantum Mechanics. I have found the cure of these epicycles also in the form of a new theory which is consistent with the Classical Physics, all these are the result of the work which has spanned for about

3 decades, and I realized that this can contribute to the progress of the Physics, therefore, decided to share through these series of books.

I am not against anybody, neither against philosophers, nor against the mainstream physicists nor against the mathematicians nor against the Physics Establishment. During my studies I have developed my own views about the working of the physical world, many of which contradicts with present accepted views of the foundation of physics and the foundations of mathematics, the main reason for the mismatch are the non-scientific steps adopted for the development of the fundamental theories. My motive is to highlight the non-scientific steps as well as to provide a purely scientific alternative theory without errors, without non-scientific steps, without paradoxes and without abstract mathematics. This may bring me in direct confrontation with the Physics Establishment, The System, but this is not the intension, my findings, conclusions, proposals are completely based on strict scientific principles, it just happens that at present they are different than the present accepted ideas.

Once being aware of these epicycles and their cures, it's hard to remain quiet, I may look insane to criticize the progress of the whole century of physics, but I believe, it worth to raise the point to the System purely on scientific and philosophical grounds and purely for scientific and philosophical intent.

I would prefer to be criticized on philosophical and scientific grounds, one of my motive is to destroy the culture of official and non-official censorship of asking genuine scientific questions, my motive is to destroy the culture of non-scientific biased personal preference of influential personalities in the development of fundamental theories of physics, which are currently treated as sacred laws of physics, my motive is also to restore the old culture of scientific inquiry, discussions and scientific methods in the light of scientific philosophy in the formation and development of fundamental theories of physics. These are the basic requirements to understand the nature of the real physical world.

I make the claim, purely on scientific grounds, that both the fundamental theories, Relativity and Quantum Mechanics are absolutely wrong, I have provided the elaborated scientific reasoning and I have proposed a complete new solution – scientific, simple, consistent, much less mathematical, mainly based on empirical experimental results. These claims should not be taken as an enemy's claims, these claims are for the sake of true physics, a course correction of century old dis-oriented efforts, I would prefer my every claim should be tested on scientific grounds to check if they pass and have some worth.

The motive is positive, constructive and scientific for the progress of Physics, since a whole century has been passed, the present situation requires to be scientifically dead critical, and I have been in these books, which may look otherwise prima facie.

6.3 IS THE AUTHOR QUALIFIED FOR THE SUBJECTS?

After making such claims, the immediate first question comes in everybody's mind is – "Is the author qualified to write on these subjects that too with a view which is completely against the mainstream?"

To answer this question it becomes necessary to look little deep what is meant by "qualification" and how this qualification is achieved and it also becomes important to know what exactly is required to write on this important subject.

In general it is supposed that a formal graduation, post-graduation or a doctorate, assures certain level of knowledge and usually it is sufficient for the requirement of industry or maintaining a laboratory or performing a cutting edge research work, or being a professor in a university, or heading a scientific project or organization, this is the system of education worldwide, the more narrow and specific the scope of study the better expert at that narrow title is expected, this system has both advantages and disadvantages, this system creates several experts in a very narrow field of subject, these experts need vast amount of time and resources and their work become so esoteric and specialized that it becomes difficult to be understood by some other expert who is in the same field but a bit different sub field, these achievements / progress of the field is validated by another process of the knowledge system – the peer review process, there are very little qualified persons available to really judge these cutting edge progress fairly and scientifically. With every new batch of researchers the knowledge keeps on building on its previous status, usually the previous work remains unquestioned. In this system, if strict scientific principles are not followed for validation, may be because of proper expertise is not available, then, there is a danger of runaway "hollow knowledge". But, this is what a practically possible feasible system can be. This system can be worse in theoretical field, where it's practical or technological application is difficult or experimental verification is in the far future or not possible at all, it becomes more prone to take the wrong path in the absence of empirical verification. The development of theoretical physics or the development of fundamental theories of physics falls in this category. I will highlight in the following books that this exactly is the case with physics. This is the reason that the higher educational certification and the rigorous peer review process is no guarantee of true scientific progress, particularly, in theoretical field, they may gradually produce systematic runaway hollow knowledge. And it becomes insurmountable if such systems have taken a wrong step and have been working for at least a century, and if it gets combined with some other non-scientific issues, it really becomes dangerous, then the system designed invested power in expert's view to take the decision and decide the direction of the progress can be incorrect and misleading. The positive strength of rigor and unbiased view of the expert of this system then becomes the most difficult hurdles to promote a true unbiased scientific view in its absence, over a period of time, the inertia and its non-scientific spell of the metamorphosed system become so powerful that they are impossible to break its non-scientific spell from within the system.

Strangely, this is the first occasion in the history of physics that this situation is faced with well-established system of knowledge and well established system of monitoring and maintaining the quality of knowledge which is guiding to a wrong path for over a century, and this is also the main reason why it has not been realized and corrected so far.

These Systems, the qualification certification and the peer review process are responsible for the progress of fundamental theories of physics by enforcing the scientific process and by maintaining the quality of the work they produce, therefore, it will not be an exaggeration to make these System responsible for the present state of the progress of physics and the fundamental theories of the physics. Now it requires to be critical about the performance of these Systems, could the System delivered so far what was expected form it? Certainly not, that's why none of the initial problems of fundamental theories have been solved, on the contrary, many of them have been suppressed and many of them are censored to be asked officially or unofficially by the System. This will keep on going with more confidence and momentum as the mega projects are producing experimental confirmations to support the fundamental theories, for example, GP-B, LIGO and Event horizon telescope etc.

Let me be dead critical, in my view, the present System cannot produce a qualified person who can have deep knowledge and understanding on the multiple core issues related to fundamental theories of Physics, the expert certification can be only for a tiny part of the subject, even if someone decided to get multiple expert certifications for several sub parts of only one part of the main theory, then the whole life of a person seems short. After a century, it becomes practically impossible for the expert system to produce an expert on all issues of any single fundamental theory. It seems the present expert or certification system along with the peer review process can only run the show in only one way, move forward towards the pseudo progress. It has become very difficult to handle the contrary views; even the views are genuine and perfectly scientific. Let us face the contradiction squarely, the System requires only the experts to address the core issues of the fundamental theories, the expert is given the status of an expert by some other experts who themselves have attained the status of experts because they have shown that they have understood the fundamental theories and have made positive contribution for the progress of the theories, then how can one expect any expert produced by the System can have views against the belief of the System itself, against the fundamental theories of physics? If he does so, then he is making himself dis-qualified, this is a real practical problem. This is the historical fact, there is a possibility that this might have not been the case, had the system entertained the contrary but scientific views, but, the fact remains, this is not the case. The contrarian views are completely suppressed, the system did not work fairly, and it is obvious it has taken a completely non-scientific biased stand in favor of the fundamental theories, it promoted false science. In my view, the System has failed miserably to guide the development of Physics in the correct direction based on scientific principles. This is also a fact that we do need some System to guide and monitor the progress of Physics and to review as necessary. In my view, the designed structure of the present System is not much problematic, the only problem is, the System lacked the true scientific rigor, it deviated from the true scientific principles to judge a claimed scientific work, this deviation is not new, it started from the beginning of the 20th century, when the foundation of this culture was laid down, the onus lies on the founders of the fundamental theories, they could have followed the example of Newton when he openly admitted that he could not solve the mystery of how exactly gravitation works at "action at a distance" and gracefully extended it for the posterity to be solved. It becomes necessary to quote Newton at this moment –

> "I have not as yet been able to discover the reason for these properties of gravity from phenomena, and I do not feign hypotheses."
>
> -Sir Isaac Newton

Even though he was the founder of the law of gravitation, the only person who understood the gravitation the most, by making this remark Newton followed the ideal condition, but hypothesis are required for the progress of physics, the only requirement is they must not be any arbitrary hypothesis; they must be based on some deep, philosophical and scientific understandings. I am sorry to say, but, the founders of the new theories of Modern Physics did not follow Newton nor the basic requirements to frame any hypothesis, rather they produced something completely non-scientific, in the name of solving the mysteries, they promoted their own non-scientific views at the expense of whole established physics, the fact is, they themselves could not understood what they have produced, this is evident from their own statements they made occasionally, they laid the foundation of false physics and with that started the culture of pseudo-science to promote their own views, not to promote the true Physics, our knowledge and expert creation system had to adopt this culture and they are still following that culture religiously – no contrary views will be entertained, how so ever scientific.

Now, is it possible for any expert, produced by the System, to challenge the fundamental theories on scientific grounds? Going from the real world facts, this seems impossible.

Let us also see, why the System behaved in such non-scientific manner? What went wrong? In my view the lack of clear understanding of the fundamental theories of physics is the main cause. As the anecdote goes, at the beginning of Relativity there were only two persons who could understood the theory of relativity, its creator and its experimental verifier and both got surprised about who the third person is, when asked by someone that is it true that there is only three persons on the Earth who could understand relativity? This may be a joke, but it highlights the true situation prevailing at that time, it highlights the image of the theory of Relativity as a very complex and very difficult to understand theory, it also highlights the genius of Einstein and a few who could understand the theory. The same goes for

Book 1: Why A New Revolution in Physics is Essential and why it is Inevitable

quantum mechanics but in a more blatant way, when Feynman made the remark "relax and enjoy and realize that no one understands quantum mechanics". In my view, these are no laughing matters, these incidences reflects the true characters of the theories and what their founders / key persons thought of the fundamental theories. When the founders and the key persons have such understanding, then, it is not hard to imagine how much their followers can understand, and which direction they will guide the expert creation system and the peer review process, and still worse, on what scientific principles they will guide? The result is in front of us, huge crises in physics, a mathematical physics with baseless equations and with a complete disconnect from the physical reality. This is not the only problem, the problem is still deeper because the frontiers of the physics today don't even realize these crises exist, they are still moving ahead like their predecessors and founders, rest assured and contented that there exists no third person.

It is very difficult for the System to roll back, even if they now become aware, it's practically very difficult to move against their own huge inertia gathered for over a period of century. In my view, this is a classic example that the world is run by human emotions, how so ever scientific methods are developed, they all will be controlled by non-scientific human emotions, this is good also, the only question is, in which direction these emotions guide the world?

It's very clear where the System created qualified experts have led us and it's also clear they will keep leading in the same direction because they are on System driven autopilot designed to run in only one direction. The end result is that, the scientific System created to guide the scientific progress of physics has failed completely, we have ended with two great epicycles – Relativity and Quantum Mechanics, and the worst part about the system is that it never tried to curb the epicycles, rather it nourished them at every possible opportunities making them ever bigger giants and still feeding them, by the recent validation of the Higgs boson and the detection of the gravitational waves by the mainstream physicists, the situation has completely gone out of control from the system to control the epicycles. This also requires to address the issues urgently.

Therefore, a revisit to the foundations of the physics is required, and an unbiased critical review is required, which in my view is almost impossible by any System created qualified expert, because the present process of making a qualified expert kills the natural instincts of un-biasness and redefines the term "scientific" which is nothing but "blind faith for the correctness of the fundamental theories". The System believes that the foundations of the fundamental theories are perfect, therefore, the revisit to the foundation of the physics is not required at all, and it only wants to move forward from the present status.

Therefore, in my view, for an unbiased critical review, someone who is not influenced by the system is required. This is the first most important qualification, and I don't have any hesitation to say that the author fulfills this qualification.

Then, what is the second most important requirement? In my view, it is the "understanding" about the physical world not the "knowledge" about the physical world. There is a vast difference between the two. The present System can only provide the knowledge about the physical world; it is worth quoting Niels Bohr, one of the key founders of the Quantum Mechanics, at this moment –

> "It is wrong to think that the task of physics is to find out how Nature is. Physics concerns what we can say about Nature".
>
> -Niels Bohr

The Quantum Physics claims it has vast amount of knowledge in non-intuitive mathematical form about the quantum world, but also accepts that they provide not a bit of understanding about how they work, surprisingly, understanding the Nature is not their goal also, as it is evident from the above statement of Niels Bohr. Let us consider another example, there is very vast amount of knowledge in mathematics and there is no limit of its further growth, as mathematical knowledge is based on arbitrary axiomatic system, theoretically unlimited sets of axioms can be defined and therefore still unlimited kind of mathematical theories and knowledge are possible, but can all these mathematical knowledge be relevant and useful in understanding of the Nature of physical reality, in other words, the understanding of Physics? No. They are absolutely irrelevant, a very tiny part of the mathematics is used in formulating the fundamental theories of physics, the majority of the mathematical knowledge is absolutely irrelevant, they don't provide any understanding neither physical nor mathematical nor philosophical, I would call it "relative knowledge" with no real values and would like to quote the great philosopher, Bertrand Russell:

Book 1: Why A New Revolution in Physics is Essential and why it is Inevitable

> "Mathematics may be defined as the subject in which we never know what we are talking about, nor whether what we are saying is true"
>
> -Bertrand Russell

Another more relevant example, just think about how many papers are published every year related to the fundamental theories of Physics, how many papers have been published from the advent of these theories so far? A vast number of papers, a vast amount of knowledge about the fundamental theories are contained in these papers, could they succeed to provide any further understanding of the fundamental theories in a true scientific sense? I doubt. Now, the question arises, why so many papers with very vast amount of knowledge failed to provide any further physical underrating about the fundamental theories? Because, all the papers are based on the knowledge of previous papers not on the understanding of the problem or the subject, it is the helplessness of the present system, it may be because of many practical constraints also, like limited resources, time, expertise, money etc. that the knowledge of previous paper is assumed correct and further knowledge is supposed to be built on it.

Understanding and Knowledge are two entirely different aspect of any system, one may lead to another, but it's not always the case, one may mislead another also and this seems the case exactly in physics, knowledge misled the understanding of the physical reality. At present the knowledge and understanding about the fundamental theories of physics are in opposite tracks with knowledge dominating and dragging the howsoever little understanding to convert into knowledge.

See the contradiction, on papers, there is progress of knowledge, a lot of progress that they are matching with high precision with the results of mega experiments, but still missing the understanding, how it all works!

In my view, the first true understanding about the nature of physical reality comes from an individual effort guided by intense passion, purely a conscious human affair; a System cannot achieve the first understanding. The first understanding is only achieved and shared by a human mind, once this happens the first understanding becomes knowledge to be taught by the System. A System can only provide knowledge of previous understandings; mind has to make an effort to develop the understanding from the knowledge provided by the System. A System do not have the capability in itself to provide a new understanding from the knowledge it contains, so far, it's the exclusive domain of only the human mind. But the precautions are necessary, if the true and rigorous scientific principles are not followed then this may prove disastrous, the first understanding may be incorrect but still pass the scientific tests because of laxity of it and become a scientific knowledge, further built up of knowledge may create an epicycle, as we are facing today, the two giant epicycles.

Therefore, in my view, the System is unable to produce an unbiased expert with true scientific understanding, who is able to perform an unbiased, scientific inquiry of the fundamental theories of physics.

There may be many individuals, who can qualify for performing an unbiased true scientific inquiry of the fundamental theories of Physics in the whole spectrum, but I have never come across anyone so far, in my view, in the beginning Einstein himself qualified as a person to raise questions on the foundation of the Quantum Mechanics on scientific and philosophical grounds, he attacked by EPR paradox, but he failed to make a mark, in recent years, in my view, Sir Roger Penrose qualifies to certain extent but the System remained unmoved and indifferent from his precious views and works.

In my view, as mentioned above, the author qualifies the first criteria by blatant, healthy unbiased, scientific criticism of the System, and the second criteria become a posterior for the author. The author has already produced the results. The formal qualification of the author becomes important and necessary, as per the present System, to select the author to perform some future task, but this is not the case here, the author has already completed the task, therefore he should be judged on the basis of his work, at this stage any kind of qualifying criteria of the author becomes irrelevant.

The System is not bad, the culture they are forced to follow is. On purely scientific grounds, anybody can put his views, therefore, I offer my view that my these books give an opportunity to the System to introspect and review the traditional non-scientific culture being followed for over a century, and adopt the scientific culture for true scientific progress of Physics. Some improvement of the System is required, in my view; the System needs more transparency and easy access to the raw data of the fundamental experiments, and an improved process to consider the views of non-mainstream individuals. I will go through the System at proper time and will submit some papers, because the System is required to scientifically manage the scientific progress of Physics.

If there is some problem in the System which has become very powerful, then it cannot be corrected by being a part of the System, an outsider is required who can provide a fresh unbiased view. This also creates problem on scientific and non-scientific grounds, but I believe, sooner or later the true scientific ideas will be accepted because there is no other option, stagnation of theoretical progress, hollowness of non-scientific understanding and technological requirements will demand for true scientific understanding and knowledge.

Therefore, it seems, the author fulfills the basic requirements at least to write on the subject, and is ready to be criticized on scientific and philosophical grounds about his views.

6.4 WHY SHOULD ANYBODY CARE FOR A NON-MAINSTREAM AUTHOR'S VIEWS AND CLAIMS?

But why should someone care for the author's views? Who is coming from nowhere, never was in the mainstream! The author's own answer to this question is – Nobody should, if the views are not based on scientific principles.

One of the main reason why somebody should care for author's view and claims is, the errors and the critical non-scientific steps in the formation of the fundamental theories are highlighted and made absolutely obvious to the extent that they cannot be ignored now. It's like $2 + 3 = 6$, nobody can ignore such obvious error. The error becomes less obvious if I write $2x + 3y = 6$, where x and y are positive integers and x is equal to y, and the value of y is 1.

Another reason is, the new ideas are completely based on empirical experimental findings, every idea and argument is based on or supported by already existing experimental facts.

One more reason is, because the author is not proposing any single or unimportant or marginally important ideas which are remotely connected with the core issues of physics and with or without these modifications the core physics remains not much affected. The author is claiming and proposing a whole spectrum of ideas on every fundamentally important issues from deep buried Classical Physics issues to the presently accepted Modern Physics issues to the extent that both the fundamental theories have to be completely replaced by a whole new theory, anybody can pick up any of the claim and idea and criticize it, check it if it is based on scientific principle and have any experimental support. If they pass, then, it won't be easy to even neglect them; they will have to be taken seriously.

Another reason is, because these are scientific ideas in true sense, they may come from anybody, so far I have not encountered anybody presenting these ideas, it just happened that these are my ideas, and I did not came across these ideas by chance, rather these are the result of consistent effort of about 3 decades. My ideas are completely open anybody can cast any doubts and can challenge them, they are open to any philosophical and scientific criticism unlike the present situations where there are many questions which are un officially banned from asking because they highlight the weakness of the theories, they are uncomfortable because nobody knows the answer, they do not have scientific explanations.

6.5 ARE THE MAIN IDEAS / THEORIES ORIGINAL?

Absolutely original. Never came across any such ideas else this revolution might have taken place earlier.

6.6 THE SOURCE OF AUTHORS IDEAS AND THEORIES

Scientific analysis of the present theories produced problems and only problems. The solution of problems, new ideas and new theories are the result of author's own solo efforts.

7
ARE THESE BOOKS BASED ON SCIENTIFIC PRINCIPLES?

Absolutely they are. My every argument and every new idea are completely based on empirical data produced by the established scientific experiments performed from pre Newtonian to modern era. The most important part of any experiment is the interpretation of the results, both Relativity and Quantum Mechanics failed in this part because for some reason they followed non-scientific interpretation which resulted in their coming catastrophe, being seen this happening and being aware of this, every care is taken to follow the true scientific principles.

8
WHAT ARE SCIENTIFIC PRINCIPLES?

This not as easy as it seems prima facie to explain, my view is different than the present accepted views, I will explain my view in short. Scientific principles are the behavior or characteristics of the Nature, the physical reality, at deeper level which are formulated as principles, usually they do not change for the same circumstances, these principles are independent of space and time, they produce the same outcome under the same conditions everywhere and every time. Scientific principles are independent of the observers and measurements. My stand of scientific principle is consistent with the Classical Physics but it is completely inconsistent with the Modern Physics - Relativity and Quantum Mechanics. Relativity and Quantum Mechanics are observer dependent theories, because of this they do not qualify as a true scientific theory in my view.

The tussle between the classical and modern view of scientific principles is discussed in the coming books, and presented in a simple language, the reader will himself be able to judge which view is correct.

9
WHAT ARE THE MAIN CLAIMS OF THE AUTHOR?

I claim that both the fundamental theories of Modern Physics – General Relativity (Including Special Relativity) and Quantum Mechanics are absolutely wrong, they are parallels of epicycles of Classical Physics; I further claim that I have found the cure of these epicycles and they should be replaced by a new theory based on new incarnation of the Classical Physics –The Quassical Physics.

It took about 5 books to highlight the failure of the Modern Physics, the failure is proved on philosophical, conceptual, scientific and mathematical grounds.

10
WHY SHOULDN'T THE CLAIMS OF THE AUTHOR BE REJECTED?

Prima facie, what I claimed here seems impossible. Let us explore further:

The first unbelievable reason is, how it is possible that the scientific work and progress of the whole century can go wrong, when we have well developed knowledge development system in place in the form of dedicated courses in universities and research laboratories which produce highly dedicated and extremely focused experts for a narrow segment of a subject for the progress of physics? Who also design and perform sophisticated experiments to check the theories often.

The answer lies in the nature of the theories and the non-scientific culture it is forced to adopt by the founders of the theories, both the theories are highly mathematical, as claimed, they provide mathematical knowledge but fail to provide any physical understanding, it's very easy to go wrong if the understanding is missing, this issue is discussed in details under- "Is the author qualified for the subject?"

The second unbelievable reason is how it is possible that the scientific work and progress of the whole century can go wrong, when we have well developed peer review process for guiding and maintaining the quality and progress of physics? Where every well-established scientific journal thoroughly judges every paper it publishes, through the very rigorous peer review process, how it is possible even a single peer reviewed paper can go wrong?

The answer lies again in the nature of the theories and the non-scientific culture it is forced to follow by the founders of the theories, because of highly mathematical nature of the theories, proper physical understanding of the theories are missing, therefore, in lack of proper scientific understanding of the theory, nobody can dare to entertain any contrarian views on these theories, the system becomes incapable for deciding on their own, and because of the forced culture, it just promotes the views the founders of the theories have shown, the same goes for the experimental results, the experiments don't lie, they reflect the physical reality but there is another very important step between the experimental result and the understanding – the interpretation of the experimental result. This acts as a very important tool, this explains why an experiment produces result precisely matching with the prediction of a theory, which are an incorrect and failed theories in a true scientific sense, for example GP-B, LHC Higgs Boson, LIGO, event horizon telescope etc. Another reason could be nobody might have produced a clear failure of the present theories on many fronts as well as nobody might have produced alternate superior scientific theories which cannot be ignored. This is going to happen this time.

The third unbelievable reason is how things can go wrong in ultra-fast communication modern day super computer era? Where things can be predicted and proved by simulations! For instance we have the super computer simulations of colliding black holes and generation of gravitational waves as well as the simulation of creation of Higgs boson and finally they were experimentally found exactly the same as their supercomputer simulation shows! Then why the modern day powerful supercomputers couldn't spot these epicycles which kept developing for a whole century?

The answer is very simple, so far any super computer is not a substitute for human mind, technology can create an artificial brain but cannot infuse the mind in it, at present we are only in the very beginning to start to understand what is mind and how mind works. The artificial brain is based on certain unconscious algorithms, which are nothing but a complex deterministic programing using copious amount of data to produce certain outcome, every outcome is perfectly deterministic, decided by the human mind in advance, through mechanical programming, the level of complexity creates an illusion of mind, but the fact remains that they are nothing but perfectly obedient number crunching machines performing mindless deterministic logical operations they are asked to perform at mind blowing speed with complete lack of any understanding. Since they don't have any capability to understand, they cannot create any new or first understanding. Any first understanding requires a biological mind; human mind seems very advanced because it is the property of the most advanced brain. Any supercomputer generates the output what the human mind teaches it, if so far, the human mind itself has not developed the first understanding about an issue or a subject then it cannot teach, through programming, the same to any supercomputer. That's why a supercomputer simulation produces only what is expected, what the human mind understands and teaches it, and therefore, it has so far failed to spot any epicycle. When the first understanding is available, and the supercomputers are taught about the epicycles then the supercomputers will spot the epicycles. In the same lines, it is also not surprising why we see the claims of the experimental results matching precisely the predictions of failed theories.

And the fourth most unbelievable reason is - how it is possible that a completely isolated person, never in the mainstream physics can ever claim the main fundamental theories of modern physics are epicycles and even more unbelievable is his claim to have fixed the epicycles by an entirely new physics which is intuitive and consistent with classical physics?

In present situation, not being in the mainstream proved the greatest advantage, need not be worried what the views of the experts of the system are, choose to create absolute freedom, have absolute freedom by being isolated, which is the very first requirement for true scientific progress. Started from the beginning, approached from the first principle, guided by unbiased natural instincts and focused on developing understanding, devoted a lot of time, involve for about three decades, completely involved for past three years. Gone through the fundamentals of physics, mathematics and philosophy, as well as, the history of physics and mathematics. The findings are unbelievable, shocked to realize how deep and how big the crises of the physics have become. Dared to solve the mysteries and really did so. This was a purely personal quest, which I have already achieved beyond my own expectation and satisfaction. Now, I feel It deserves to be shared with the world because it worth.

I accept that my these claims are really unbelievable, and I would prefer nobody should believe these claims on face value, just because something is said by someone, it should not be accepted as well as it should not be rejected also in scientific spirit. I would prefer that every claim should be questioned on scientific grounds then only it should be

Book 1: Why A New Revolution in Physics is Essential and why it is Inevitable

decided if the claims sustain or fail. But, the big question is why should someone spend their precious time on reading these books? The answer is, because physics is going through big crises, the crises become more dangerous because they have been ignored for over a century, many of the top mainstream physicists are not even aware that there are crises in physics, some are aware but don't accept officially, some physicist are aware but they just ignore it, a still very few tried to drop a warning and raised their voice for some cases but the momentum of the developing century old epicycles are so great that their voices are doomed, in this regard I would like to mention about Sir Roger Penrose, he is very clear about the emperor's new clothes, but nobody is paying any heed to his warning and the procession is still going on, but, I am confident that these 10 books of the series "A New Revolution in Physics" will certainly stop the procession, sooner or later, because, there is no other alternative, the epicycles must be realizes and cured, these are the only cures of the epicycles because the cures are truly scientific without any paradoxes and is consistent with the experimental based grand old Classical Physics, the cure of the epicycles will have to be accepted, the cures are really like the heliocentric solution of the geocentric epicycles, it happens only once and thereafter no turning back, because it represent the physical reality, it cannot be undone or interpreted in another way, the same parallel is here in the new theory of physics –the Quassical Physics, once revealed it can't be neglected, it will force to be accepted by its sheer content of the physical reality, like the heliocentric theory, timeless forever..

11
WHAT IS "A NEW REVOLUTION IN PHYSICS"?

A New revolution in Physics is an attempt to restore the true sense of physics, like it was in the Classical Physics, when to understand any physical concept mathematics was not required at all, mathematics was required only to calculate the magnitude of the physical quantities involved, mathematics had nothing to do with the understanding of the physical concept involved, I am very much aware that my this view is contrary to the great Galileo, but I have my reasons for the same. For example, let us take gravitation, gravitation is a physical concept which tells that a massive body creates a sphere of influence around itself such that any other massive body comes within this sphere of influence of the first body, it experiences a force of attraction towards the first massive body, experimentally it has been found that this force of attraction is proportional to the mass of both the bodies and inversely proportional to the square of the distance between them. This is the empirical fact of the Natural world of physical reality, this happens every time, this happens everywhere, with every mass small or gigantic, for everyone looking or not looking, for everything moving or not moving, for everything live or dead, the results remains the same for the same conditions. This is the property of the real physical world we live in; it exists with some definite properties independent of any living thing including the humans. Physics starts with the assumption of the existence of the real physical world, and its aim is to find out the properties of this real physical world, the assumption of the existence is not arbitrary, the assumption is based on the direct experience of

Book 1: Why A New Revolution in Physics is Essential and why it is Inevitable

human mind with long term constant and consistent interactions with the real physical world as well as it is also based on the observation of constant and consistent behavior of non-living bodies across the spectrum of distance, size and time. This is the very basic philosophy of physics. The Classical Physics followed this philosophy and has produced important fundamental concepts of physics and thus laid the solid foundation, it should be said solid foundation because the fundamental laws discovered by the Classical Physics are mainly based on empirical data and its interpretations was governed by true scientific principles and the above described philosophy. The Classical Physical did its best with the minimal state of technology available with it, and the founders of the Classical Physics had the passion for real physics, passion to discover the laws of the Nature, the laws of the real physical world. They were respecting the Nature and were aware of their limitations that they cannot frame the laws of the Nature, they can only discover them.

Unfortunately, in the 19th century the philosophy of physics started to change and it took the center stage in the beginning of the 20th century in the form of Einstein's 1905 paper "On the electrodynamics of moving bodies" which formed the basis for Special theory of Relativity. Special theory of Relativity is mainly based on two postulates, in my view, both the postulates are giant deviation from the classical philosophy, it will not be an exaggeration if I call them anti-scientific and anti-philosophical, by virtue of giving the status of postulates, it is barred from the scientific and philosophical scrutiny and discussions, I am still amazed why this was scientifically and philosophically not discussed and calmly accepted by the other great physicists of that time! These postulates were not just new few lines added to the physics, they started the dangerous culture in physics and very silently, completely killed the Grand Old Classical Physics, it changed many of its basic concepts like space, time, length, mass, energy, etc. In the noise of revolutionary concepts, I think nobody realized the damage it inflicted on the Classical Physics, in my view, even today, majority of mainstream physicists are completely unaware of it. This cultured blossomed and created the General theory of Relativity, by snatching the gem, the Newton's theory of gravitation, from the Classical Physics, General Relativity reduced the Classical Physics to the status of approximate physics, the respect for the Nature was at the lowest, the remark appeared "...then I would be sorry for the God." The philosophy of physics has taken a complete u tern, now a theoretical physicist is framing some laws and expecting the God, the Nature, to follow them!!! And to the surprise of the God, almost every physicist, thereafter, are expecting the same.

The culture continued and created the Quantum Mechanics, the founders of the Quantum Mechanics deeply inherited this culture and created a new high, they denied the existence of the God himself, Quantum Mechanics denied the existence of the real physical world itself! It would be very interesting to know that when Einstein had debate with Heisenberg, founder of one of the pillar of Quantum Mechanics, the uncertainty principle, Einstein was not happy with Heisenberg's work on foundation of Quantum Mechanics, he objected about the abstractness of the concepts, to which Heisenberg replied that he is simply following Einstein what he did in Relativity

Einstein was very much worried about the probabilistic foundation of quantum mechanics, Einstein told Max Born

> Quantum theory yields much, but it hardly brings us close to the Old One's secrets. I, in any case, am convinced He does not play dice with the universe."
>
> -Einstein to Max Born

Although in his later years Einstein had real concerns about the soundness of the foundation of physics, and he was the most important and influential personality to do so, to fight against the founders of Quantum Mechanics to save the heart of the Classical Physics -the causality or determinism, but, he could not succeed, the culture Einstein had started had become so arrogant that it didn't spare his own creator, Einstein was reaping what he had sowed during the foundation of Relativity, he became the victim of his own creation, although Einstein had remarked in his later years "... God does not play dice." he had already ordered the God to obey the rules he himself had created long back. After the debates were over, Einstein was considered defeated and Niels Bohr and company were considered the winner, they used this victory to push the culture to new heights, that's why we have today all sort of non-scientific and non-philosophical theories mushrooming around. It must be noted that, the exchange of words are extremely important, as they show the true nature of what was going on actually behind the scene unofficially, Einstein was unhappy at the core about the true status of the physics on personal level, but officially Einstein also did not accept his mistakes in Relativity which personally he seemed to have convinced in his later years. Bohr, Heisenberg and others were adamant, they never seemed to be worried about the correctness of the foundation of the physics; it seemed they were worried of, and made every attempt by using their influences in every respect to promote their views, therefore, after the defeat of Einstein, officially everything moved as nothing has happened, and is still moving on. The non-scientific and non-philosophical culture is reaching new levels, now it seems they have convinced the God.

I may be severely criticized for my views about the founders, great minds and heroes of the Modern Physics, but it's over a century now, and somebody is required to tell the truth, somebody is required to make a fair comment on the Emperor's new clothes. Somebody is required to dare against the Giants, the Culture and the System. I have made critical and fair comments not only on the heroes of the physics but also on the other physicists who are running the show after the original heroes have gone, the authorities of the physics and the peer review system, just the necessary comments urgently needed to be made.

The New revolution in Physics is my attempt to stop this culture, by highlighting the errors, mistakes and non-scientific steps adopted in the development of physics for over a century, I have found the errors, mistakes, the reasons of the above discomfort, and I have achieve the solutions also, I know it is unbelievable, but till its unseen.

If the mainstream physicists and the peer review system scientifically and critically review my work in an unbiased way, a true physics can be established with all the errors of the century corrected, with a resurrected Classical Physics, clean, consistent, scientific, much less mathematical and a bit more philosophical – The Quassical Physics, else power rules, it may cause some delay, but I believe it cannot be stopped..

12
WHY THE NEW REVOLUTION IN PHYSICS IS ESSENTIAL?

New revolution in physics will only be essential when the current theories fail to fulfill their due responsibilities, they fail to provide the required understandings of Physics, they are unable to explain the observed phenomenon of the real physical world, but, so far every mainstream physicists claim that the present theories of physics mainly Relativity and Quantum Mechanics are the great achievements of mankind which precisely explain the observed phenomenon of the real physical world, they provide lots of experimental "proofs" in support of the current theories, they produce lots of very complex, long but beautiful mathematical equations to prove their point, they even claim that not a single experiment or observation has so far found contradicting these theories! Everything is so fine, except some minor problems. Then where is the problem? How can I be so insane, on the contrary, to claim that new revolution in physics is essential?

The answer lies in details, in the fine prints. It is required to look beyond the selective claims of the success; even it is required to scientifically analyze the claimed success, which has been done in the following books. Both the well accepted fundamental theories Relativity and Quantum Mechanics have many common characteristics which are non-scientific. New revolution in Physics is essential because of the following reasons regarding both the theories:
- A. The basic philosophy is non-physical and non-scientific
- B. These theories are based on a person's or a groups' intuition or believes without scientific scrutiny.

C. Provides pseudo solution, when unable to explain, accepted the problem itself as postulate
D. Use abstract non-physical concepts.
E. Are based on abstract mathematics which lack physical understanding
F. Are haunted by appearance of infinities, which is the sign of failure of the theories.
G. Give pseudo-scientific answers to fundamental questions
H. Predictions do not match with real physical world
I. Claims of solving problems of real physical world are wrong
J. All experimental "proofs" are under cloud

Let us explain one by one in short, they are explained in details in the other corresponding books of this series.

A. **The Basic Philosophy Is Non-Physical And Non-Scientific:** It will be very strange to know that the basic philosophy followed in the creation of theory of Relativity and Quantum Mechanics are non-scientific, this is one of the most important aspect which has been ignored completely, it seems majority of physicist are not even aware of this most important fundamental problem of the foundation of physics. Physics must be driven by scientific philosophy, but it seems that the modern physicist have completely ignored the necessity of philosophy in physics and they have also ignored that the founder physicists had followed non-scientific philosophy of physics and therefore, the present physics is in this abandoned status, clueless about where to go and how to go and what to do.

It will be surprising to know that the founders of the modern fundamental theories, Einstein, Niels Bohr, Heisenberg and company all were deeply impressed by the Immanuel Kant's philosophy of science and they based their theories on Kant's philosophy, but what is Kant's philosophy of science?

In short, Kant's philosophy of science is only for the sake of philosophy, it has nothing to do with physics, but it was applied in the fundamental theories of Physics by our founders and we have these non-scientific fundamental theories, bizarre, disconnected from the physical reality. Kant's philosophy of science is based on the primacy of consciousness not on the primacy of the existence, which in simple language means that the rejection of the existence of the real physical world! The physical world doesn't exist at all, what we sense from our sensory organs are only the appearances which is created by the consciousness, with this philosophy; he further invaded the realm of Physics and claimed that, the aim of physics is only the mathematical description of these appearances. Einstein, Niels Bohr and company implemented this Kantian philosophy in the formulation of fundamental theories of Physics and killed the heart of true Physics, The Classical Physics.

It is not important whether they have openly accepted that they followed the Kantian philosophy, but, the imprint is dead obvious, and, this is also a fact that majority of mainstream physicists who are staunch supporter of Relativity and Quantum Mechanics, do not know that these fundamental theories of Physics are based on the philosophy in which the real physical world doesn't exist at all! I reject this philosophy altogether, and, every physicists must also reject this philosophy as the philosophy of Physics. In my view, it is more than insane to accept the Kantian philosophy for Physics. Let me make it clear, I am not against the whole philosophy, I don't bother a bit if such philosophy or even its higher version according to which the philosopher himself don't exist, are discussed, debated and somehow implemented, but within the sphere of philosophy itself, the moment it starts to invade the realm of Physics, I have the concerns, and everybody must be concerned, including the mainstream physicists.

The fact is, all those philosophers who do not believe in the existence of the real physical world, seem to follow the laws of the real physical world, they go to the doctor when they become ill and take the medicines which is developed by the real physical world, they are never found walking in the middle of the road, believing their philosophy that all the vehicles are only appearances created by the mind and the impact cannot harm them, they do not act what they believe, they are aware the impact of the appearance of the vehicles will definitely harm them and they will again need the help of the real physical world, because this is the EMPIRICAL fact, in a physicist's language this is called the property of a real physical world, it is the manifestation of the existence of the real physical world, empirical knowledge is the knowledge of the real physical world.

This is the fact that these philosophers do not follow their own philosophy in practice, because they cannot ignore the empirical facts, but, they have every right to have their opinion and philosophy, may be after a long, long, time, in the end, they may prove right, because we don't know how the brain works and how the mind becomes aware of the physical reality, also because the very nature of philosophy allows them to believe so, they are not bounded to the empirical facts only to frame their philosophy, therefore, let the philosophers be worried about the non-existence of the real physical world. But, a physicist is absolutely bounded to the empirical facts, the empirical facts must be the basis of all the theories a physicist creates, he may be guided by some scientific philosophy which is consistence with the existence of the real physical world, but, the finality is the empirical facts only, because the aim of the physics is to know about the real physical world and to understand its behavior. In my view, the scientific and critical view should be, if any physicist does not believe in this aim, then, he is not a physicist, he belongs to a higher category league of philosophers and therefore, their opinions and theories should not be taken seriously by the real, hard-core, empirical physicists.

Physics is the study of the real physical world, period. The real physical world exists independent of mind, period. The Nature has made it that a human mind takes pleasure to discover the secretes of the Nature, it may be a philosophical game played by the Nature, but it worth for the existence of the whole human race, deep knowledge and understanding of the Nature is necessary to protect his own existence at present and in future. Physics is useful; it's a necessity for survival. It's not a futile armchair exercise. Let us bother about the present, let us bother about understanding the real physical world in present and near future, let us leave the distant future for the philosophers.

In my view, the role of observer in the theory of relativity and the role of measurement in quantum mechanics are the direct evidences of the influence of Kantian philosophy, whose main premise is that the real physical world does not exist, then, how can one expect that these theories, Relativity and Quantum Mechanics, which are based on this philosophy, will reveal and predict the true nature of the real physical world? Let me be dead critical, it's pure insanity, and we have been appreciating these insanity as the triumph of the human knowledge, the greatest achievement of the human mind! In my view, this single reason is more than enough for urgent and deep introspection for the mainstream physicists; it's enough to initiate a new revolution in Physics, to clean the Physics of all such theories which is based on non-existence of the real physical world.

B.   These Theories Are Based On A Person's Or A Groups' Intuition

Or Beliefs Without Scientific Scrutiny: This becomes the natural outcome when the Kantian philosophy is followed, the other premise of the Kantian philosophy says that the aim of physics is the mathematical description of the appearances, therefore, in Special theory of Relativity and the General theory of Relativity, Einstein took the lead to describe the appearances in mathematical forms by adopting the mathematical Lorentz's transformation and by creating mathematically highly complex Field equations of General theory of Relativity, Bohr described by stationary quantized orbits, Heisenberg described by uncertainty principle and Schrodinger described mathematically by the Schrodinger's equations. All these mathematical descriptions claim that they are deeply connected with the real physical world and represent the physical reality, but the fact remains, they do not represent the physical reality, these are just opinions of individuals, they only represent the appearances in the Kantian sense and have never gone a true scientific scrutiny. This is the reason why we have very exciting physics today, which is very entertaining and allows all sort of wonders to happen. The scientific reason of my claim that these do not represent the physical reality is discussed in details in the other books of the series.

C. Provides Pseudo Solution, When Unable to Explain, Accepted the Problem Itself as Postulate: This is very strange, both Einstein and Bohr did this. When Einstein was unable to provide the scientific explanation of the Michelson Morley experiment, he accepted the problem itself as postulate, the seemed constancy of speed of light in his Special theory of Relativity, he did the same in the General theory of relativity when he could not provide any explanation for the equivalence of inertial mass and the gravitational mass, he assumed this as the law of Nature, a postulate as the equivalence principle. Niels Bohr also did the same, when he could not provide the explanations of the stability of atom, he assumed the problems, the empirical findings, as the postulates of his theory – the quantized stationary orbits, Heisenberg did the same, de Broglie did the same, Schrodinger did the same, Pauli did the same, Born did the same, Dirac did the same and so on. How could they do this? Because the Kantian philosophy allows this, else a real physicist cannot take such non-scientific steps. This I call the "pseudo solution", the problem still exists without any explanation but by assuming it as a postulate makes it unquestionable and removes it from the list of problems, and this is very dangerous because now this non-scientific postulate is ready to build a theory upon itself, ready to promote a pseudo-science. The result is in front of us – the two giant epicycles.

D. Use Abstract Nonphysical Concepts: Once the problem is accepted as a postulate and a new theory is built upon it, then how this theory is going to explain the physical reality? It may crawl a step or two, and will finally fail, but is expected to explain a vast amount of observations then how it will do? It will invent esoteric highly mathematical, non-physical concepts which are beyond the understanding of normal person or normal physicists, only the experts could understand. The examples are the 4-dimensional space-time, the curvature of 4-dimensional space-time, the wave function, the orbitals etc.

E. Both the Theories Are Based on Abstract Mathematics Which Lack Physical Understanding: If a theory is based on wrong postulates then only abstract mathematics can help it to look like a scientific theory. Relativity used non-Euclidean geometries and Quantum mechanics used Complex numbers. I have explained in my other books of this series both these branches of mathematics have nothing to do with the real physical world.

F. Both The Theories Are Haunted By Appearance Of Infinities, Which Is The Sign Of Failure Of The Theories: The field equations of General Relativity is haunted by the appearance of infinity, but even the obvious failure is viewed as the greatness of the theory, the strange property of the Black holes which is not completely understood by the physicists themselves. The most surprising thing is that, the relativistic infinity comes in different shapes and sizes! The quantum field theory is greatly haunted by the appearances of infinities; its major efforts are how to curb the infinities. The fact is, infinity of any kind is a purely mathematical concept, it cannot exist in a real physical world because of its own property itself, that's why they are the sure sign of failure of a theory, but, present theories predict large number of infinities, there exist at least one infinity at the center of every Black hole and there could be more than billions of black holes in the visible universe.

G. Both the Theories Give Pseudo-Scientific Answers to Fundamental Questions: Just ask how time can be considered as an equivalent dimension of space, how to resolve the twin paradox? How a 4-dimensional space-time gets curved in 4-dimensional space-time itself? How a curved space-time creates a force on a mass resting on the Earth's surface, as gravity is not a force it's only a geometric property according to general relativity. What exactly is an orbital produced by the solution of the Schrodinger's equation? In what form exactly an electron lives inside an atom? How the quantum jump takes place? The chances are, one may get very complex answers ending with the sentences – the world is like this only, the equations are telling this, asking more will be too much.

H. Many Predictions Do Not Match with Real Physical World: Prediction of the Special theory of Relativity – mass swelling, which violates the most basic principle of conservation of energy, and even Einstein refused to accept it. Time dilation creates the true paradox, the twin paradox, it has survived all possible explanations by the relativists, length contraction creates train and tunnel paradox which cannot be explained. It is very difficult to know how the field equation of the General theory of Relativity can make any prediction; one of the predictions is black hole, which contains infinity at its center. Another prediction is gravitational waves, which has already been detected and has made a recent Nobel Prize, but, how a curved space-time, which is geometry, creates a gravitational wave which propagates at the speed of light is still scientifically an open question. The Schrodinger's equation which is based on the postulates of the Quantum Mechanics, predicts only 3 quantum numbers and therefore predicts half the number of electrons than an atom actually has!, this is a big mismatch, 100% out, between the prediction of the quantum mechanics and the real physical world, to make the complete prediction a fourth quantum number is required which is called the spin quantum number, but, this is an empirical finding, it has nothing to do with the postulates and theory of quantum mechanics. But the claims are absolutely contrary.

Book 1: Why A New Revolution in Physics is Essential and why it is Inevitable

I. Claims of Solving Problems of Real Physical World are Wrong: gravitation is not the curvature of space-time and quantum mechanics does not explain the empirical periodic tables, these two issues are explained in details in my books – "The Collapse of the General theory of Relativity" and "The Collapse of the Quantum Physics".

J. All the Experimental "Proofs" of both the Theories are under Clouds: this seems impossible, but let us believe in scientific principles, the foundations of these theories are non-physical and non-scientific, they cannot explain the real physical world, every major experiment is discussed in the respective books.

13
WHAT IS THE ACCEPTED MAIN PROBLEM OF THE FUNDAMENTAL THEORIES OF MODERN PHYSICS?

In my view, one of the main problems of Modern Physics accepted by the mainstream physicists is the incompatibility of the two main theories of physics, one theory is General theory of Relativity, which mainly deals with gravitation, space, time, simultaneity, light, velocity, energy, in a sense the objects which are larger than used in daily life, till galactic scale and probably beyond, but the effect of Relativity cannot be experienced in daily life activities because their effects are far too small to be noticed and measured, the other theory is Quantum Mechanics, the theory of Quantum Mechanics mainly deals with the smaller objects generally atoms and molecules and smaller than these, in other words smaller objects which we cannot observe in our daily life. Here also the effect of the quantum theories cannot be seen and measured in daily life because the effects are visible and measurable for very small objects molecules or smaller than those. Isn't it strange that the effects of the two main theories of Modern Physics cannot be observed in daily life activities, matter of normal sizes and normal velocities? The effect of one theory becomes observable with matter of large size or very high velocities and the effects of the other theory becomes observable with matter of very small size, then, which theory can explain the observations of daily life macroscopic activities? They are the theories of the Classical Physics, which is mainly based on empirical findings. The Classical Physics explains the vast amount of the observational activities of the real physical world, but the strange thing is that, the Classical Physics has been declared failed by the Modern Physics because it was claimed it could not explain some of the observational phenomena of the daily life, which led the birth of Relativity and Quantum Mechanics.

Therefore, the vast majority of the observational phenomenon of the real physical world are explained NOT by the two main theories of physics but by the declared failed theory of Classical Physics. It is further claimed that the Classical Physics is only approximation of the main theories which seems to work only within certain limiting conditions and it is not accurate, it only provides an approximate results. This is the accepted status of the physics at present. This is really strange.

Now, as mentioned above, one of the main problems of the current physics is that both the theories, Relativity and Quantum Mechanics, are incompatible to each other, so what? Let them be incompatible to each other, what's the problem if they are working within their respective domains perfectly?

Yes, this incompatibility can be conveniently ignored, and it is ignored by majority of the mainstream physicists, they only want to talk about the claimed success of both the theories, they produce experimental evidences in support of these theories, the latest discoveries related to relativity are measurements of frame dragging and geodetic effect by Gravity probe B which is matching with good accuracy with the prediction of general relativity, the detection of gravitational waves by LIGO, the first ever picture of a Black hole by Event Horizon telescope and for quantum physics the discovery of the God particle or the Higgs Boson by LHC. These are accepted by the mainstream physicists; therefore, these are hard-core experimental confirmations of these theories. It is believed that these theories are correct and somehow they will be reconciled in future by some mathematical techniques.

Now, as mentioned above, since we have the experimental evidences which matches with the predictions of these theories precisely, it can be concluded that both the theories are so far correct in their respective domains else these experiments might have failed, they could not have produced so precise confirmation of the predictions of the theories, now it is expected with greater degree of confidence that both the theories will be reconciled, but It must be noted that all the efforts made so far to reconcile these theories are proved in vain, and it seems that there is almost no hope that they can be reconciled on theoretical grounds in their present forms as they are, the reason for this pessimism is the mathematical frameworks of both the theories. The mathematical framework of both the theories are completely different, theoretically it seems impossible to reconcile them unless either any one or both the theories are modified drastically, compromising their identity greatly, but in that case what will happen to the firm experimental evidences which have been produced in support of these theories, as mentioned above? Is it possible to go back and modify the theories in such a way that they can be reconciled but still maintain their predictions the same which have been proved experimentally? The optimistic may argue why can't there be any possibility? There is always a possibility for anything, howsoever small, why not for this case? In my view, the negative answer lies somewhere else, not in this probability game.

Let us explore why this incompatibility problem should not be ignored altogether and why there are some physicists who are taking the pain to solve this? In my view, as the experimental evidences are continuously mounting in support of these theories, it's only a matter of time before the incompatibility problem will be completely ignored. After some more claims of experimental evidences, nobody can dare to doubt the entire mountain of claimed experimental evidences, as it has happened in the history, during the development of foundation of Quantum Mechanics, the mainstream physicists will claim and accept that the world is like this only, it has to be accepted, there is no choice left, and both the theories will keep moving in their own domains and the core issues will be forgotten, like the problem of cause of inertia, for example.

Now, what is the meaning of accepting these theories as they are today without reconciliation? This means that the real physical world is not seamless from very small to very large, in other words, the Nature, the real physical world has two different sets of rules, one set of rules for small objects and entirely different set of rules for larger objects, but the problem is, as we know any large object, how so ever large are just the collection of vast numbers of small objects, atoms and sub atomic particles, then why there exists two entirely different sets of rules? Why not the rules which are applicable for the microscopic world, atoms and subatomic world are still valid in the macroscopic world which is just collection of more number of atoms and subatomic particles, and where is the boundary of this distinction, and how the Nature keeps track of the transition from microscopic to the macroscopic world? The bigger question is – Is the Nature, the real physical world, really like this and do

Book 1: Why A New Revolution in Physics is Essential and why it is Inevitable

we understand correctly the dual rules of Nature or, is there something terribly wrong in our understanding of the rules of the Nature?

The question comes across – is the problem of incompatibility really an important and a big problem or it's just a minor problem which can easily be ignored without having much effect on the physics itself? In my view the problem of incompatibility is very important and a great problem and I suppose that the mainstream physicists also consider this as an important big problem, that's why there is some effort taking place to resolve it and some physicists are making the efforts, but it is not given due importance, sufficient effort is not made or the efforts made gone in the wrong non-scientific path.

In my view, the problem of incompatibility is the greatest problem in physics today, and it must be addressed urgently. Let me explain why I consider the problem of incompatibility the greatest problem of physics today, the reason is, it indicates something is terribly wrong, either any one of the theory is terribly wrong or both the theories are terribly wrong, It is non-philosophical and non-scientific to imagine that the Nature will discriminate between an small and a large object, the non-living, inanimate physical world will discriminate between the two objects, it is our understanding of the physical world which is responsible for the creation of these two sets of rules, these two different theories, therefore, it is more probable that our understanding is wrong. Why this incompatibility problem is urgent to resolve? because the physics today has taken the wrong path and it tries to move forward but for many decades it has become stagnant, no progress has been made on theoretical fronts in real sense, since lots of efforts are being made to push forward these theories, they are resulting in many non-scientific and nonphysical hypothesis and theories which has absolutely nothing to do with the real physical world. The efforts being made are becoming worthless.

One of the main reasons why Relativity and Quantum Physics are claimed to be successful is that both the theories are highly mathematical. It will be appropriate to quote Einstein at this moment:

> "Since the mathematicians have invaded the theory of relativity, I do not understand it myself anymore."
>
> -Albert Einstein.

This remark should not be taken for fun; this is a very serious remark which shows the frustration of a physicist for the use of mathematics in his own theory, but, mathematicians cannot be blamed for this situation; it is the physicists who should be blamed who allowed abstract mathematics in their theory without understanding their physical significance.

This is one of the hints that the theory of relativity may go wrong. Nobody understands it neither its creator, Einstein, nor the persons who supported him to use some tools in the derivation of the mathematical model –the mathematicians. Mathematicians did their work, for them they were following the laws of mathematics, they were simply applying their axiomatic mathematics, it was the job of all the physicists, including Einstein, who could not take the pain to understand its physical significance and accepted the mathematical part of the theory. Simply making the above comment cannot absolve him of his responsibilities , if he is intended to produce a theory which is supposed to explain the behavior of the physical world, he was aware that he was drafting the fundamental theory of physics, he was well aware of his responsibilities. In my view, Einstein failed to justify his roll for the responsibility he assumed himself. I would like to quote his another remark

"…then I would be sorry for the God."

-Albert Einstein.

Again, let me become dead critical, which is necessary to deal with such remark in science, particularly in Physics, this is one of the most un-scientific and irresponsible remark, in my view, any physicist has ever made. This is against the spirit of science, science particularly physics is based on experimental interaction with the physical world (someone may raise an objection if the physical world really exists? I have discussed this in detail in other book), we cannot dictate the Nature, the real physical world, how to behave, we can only observe the behavior of the real physical world and based on that we can extract some laws, the experimental results are supreme. This remark is highly un-scientific and misleading. This remark acted as a fort to protect Relativity from any scientific criticism and it is still working well. Nobody can dare to challenge a person who is challenging the God, particularly when nobody understands, even its creator, the so called scientific work for which the God is challenged. This is not science, certainly not physics. Einstein made his position, and therefore his theories unquestionable by such remarks, there are many such remarks, he created a spell, purely an un-scientific spell, the mainstream physicists are still in full grip. This situation is further worsen by the peer reviewed process of publishing the scientific works, because nobody understands relativity and they are still under the spell Einstein has created, it becomes imperative for them to keep the spell going, therefore, there is no effort seen to really understand relativity, rather, if anything or anybody even starts to suggest contrary to relativity, is systematically purged, this is a practical un-official fact. This fact cannot be accepted officially, this is how our practical world works, like politics, there are two sets of rules, one set of rules to put on papers which are official and the other set of rules to be actually followed, non-existence on paper but can be experienced directly. The whole system has become so strong, so contented, so self-referenced and so peer reviewed and so confident and so powerful that it is not possible to break the spell and see the scientific reality of Relativity from within the system. An "outsider" is needed, who so ever he/she may be. I have no hesitation to consider myself an "outsider" who is making an effort to break the spell on philosophical, conceptual, theoretical, mathematical and experimental grounds.

The same situation is in quantum physics, the only difference is, Relativity is supposed to be the conceptual brain child of a single person, Einstein, although it has many contributor mathematicians, whereas Quantum Physics is the result of several contributors, the initiator and main contributor being Niels Bohr, therefore it is Niels Bohr and company who are responsible for the debacle of Quantum Physics.

Here the failure is more obvious, need to quote the famous remark of Richard Feynman, one of the key figures in the development of foundation of Quantum Mechanics, about the Quantum Mechanics:

> "Relax and enjoy it and realize that nobody understands it"
> -Richard Feynman

Again, this is no fun, by painting it as a causal remark its gravity cannot be undermined. This is a very serious remark which is still valid today. In my next book "The Collapse of Quantum Physics" I have discussed in detail what could have made a physicist like Richard Feynman to make this comment and many things beyond that to justify the complete failure of Quantum Mechanics.

It is important to note the philosophy of the founders of the Quantum Mechanics, they believed that the aim of Quantum Mechanics is NOT to know about the world, it is only what it can be said about the world, it is in absolute contradiction with the philosophy of the Grand Old Classical Physics, the aim of the Classical Physics is to know and understand about the real physical world, for Classical Physics, there exist a real physical world independent of the human mind whether it makes any inquiry or not, but for Quantum Physics, primarily, the world doesn't exist at all, the measurement became more important, and it's the measurement which makes the world appear, it becomes even more surprising when it is claimed that, the observer is the part of the measuring system, and if you think that this is enough than it certainly not, it goes to the consciousness of the observer, it may be a human consciousness or a bacteria's consciousness, howsoever primitive, it can collapse the quantum world from the superposition states of many possibilities into the appearance of a single possibility, somehow, this behavior of the quantum world is also claimed applicable to the macroscopic world where the quantum magic don't seem to work, where the rules of the grand old Classical Physics works, the Quantum magic is summarized in a famous single sentence

"..if you are not looking at the moon it doesn't exist".

This is only one of the interpretations of the Quantum Mechanics called the Copenhagen interpretation, there are many interpretations but, here I will not go into each of the interpretation, only I will mention just another one, which definitely makes you to doubt your own existence, if you really exist? Are you the same before reading this sentence? Are you the original one or a copy of yourself? Endless questions will spring out but all these questions can be "scientifically" answered with a quantum theory known as the Many world interpretations, as per this theory, when a measurement is made then all possible outcome exist simultaneously in parallel Universes, the single measurement creates multiple parallel Universes, everything of all the created Universe remain the same the only differences are they contain they contain different possibilities of measurement. A lot of effort has already gone into this and is a very serious research topic. I don't know how many parallel Universes and how many copy of yourself I might have created by asking the question just a minute before "if you really exist?" you certainly might have thought something and would have reached into the conclusion that "yes I do exist", for Quantum Mechanics this is a measurement the consciousness has made and is capable of creating many parallel Universes each having your copy, may

Book 1: Why A New Revolution in Physics is Essential and why it is Inevitable

be in one of the parallel Universe you are still struggling how to check you exist or in another parallel Universe you are still thinking what is the meaning of existence or in still another parallel Universe you are still busy performing the check and in this process have already created many more parallel Universes. This is really a scientific serious research topics, in my view, if this is called scientific then the word "scientific" has lost its meaning, it needs to be restored, how and why we have reached into this new state is discussed in my book – The Collapse of the Quantum Physics, and how the true meaning is restored is given in my another book – The Concepts, The Dawn of a New Physics – The Quassical Physics.

It is very important to quote the other founders of the Quantum Physics also, how remote to the heart of the physics they are and also how contradictory they are from one another:

Niels Bohr comments:

> "It is wrong to think that the task of physics is to find out how Nature is. Physics concerns what we can say about Nature"
>
> -Niels Bohr

> "If quantum mechanics hasn't profoundly shocked you, you haven't understood yet."
>
> -Niels Bohr

Pauli's comment:
> "The electron has a 'two-valueness' not describable classically."
>
> -Wolfgang Pauli

Heisenberg's comment:
> "I repeated to myself again and again the question: can Nature possibly be as absurd as it seemed to us in these atomic experiments?"
>
> -Warner Heisenberg

Schrodinger's comment:
> "If we are still going to put up with these damn quantum

jumps, I am sorry that I ever had anything to do with quantum theory"
-Erwin Schrodinger to Niels Bohr

P A M Daric's comment:
"The fundamental laws necessary for the mathematical treatment of a large part of the physics and the whole of chemistry are thus completely known, and the difficulty lies only in the fact that application of these laws leads to equations that are too complex to be solved."
-P A M Daric

Richard Feynman's comment:
"Quantum mechanics describes nature as absurd from the point of view of common sense, and yet it fully agrees with experiment, so I hope you can accept nature as she is -- absurd."
-Richard Feynman

Einstein's comment:
"God does not play dice."
-Albert Einstein to Niels Bohr

Niels Bohr's comment:
"Don't ask God what to do."
- Niels Bohr to Albert Einstein

These are only some of the important comments of the founders of the development of the two currently accepted fundamental theories of the Modern Physics, which un-officially tells a lot that the development is not entirely guided by scientific principles rather it was mainly driven by personal preferred views. This is explained in more details in the respective books.

Now let us see what are the scientific basis on which these theories should be declared that they are either incomplete, misleading, incorrect or absolutely wrong.

Any theory is called a scientific theory if it has at least two properties:
    a. It explains vast amount of the observed phenomenon of the

physical world.
b. It has the predictive power which can be tested by proper experimentations.

Both the theories, Relativity and Quantum Mechanics claim to meet the above two requirements, but the fact is, they do not meet any of the above requirements to qualify as a fundamental scientific theory, the claims are in place but they do not pass scientific scrutiny, in short they are tabulated in this book, and in details they are described in the other respective books of this series..

14
WHY THE NEW REVOLUTION IN PHYSICS IS INEVITABLE?

There is very short and simple answer to this question, because, the true physics behind the Quantum Mechanics epicycle has been discovered as well as the true physics behind the Relativity epicycle has also been discovered. The cures of these epicycles, once disclosed cannot be neglected and suppressed for long, because they are completely based on empirical findings, scientific principle and on scientific philosophy.

There is a little difference between the two epicycles, prima facie, Quantum Mechanics qualifies as epicycles because at face value, as it claims and seems to explain the subatomic world but this is also not true equivalent to epicycles, epicycles really explained observed planetary motions and had the correct predicted path of any planate which can be observationally verified but failed to explain some other observations like phases of Venus, on the other hand Special Relativity and General Relativity doesn't even qualify as epicycles, Relativity is worse than epicycles, it doesn't explain a bit it claims to or it is supposed to, Relativity is only believed that it explains many things and is better than Newton's gravity, this is a false belief, this is a blind faith, an absolute scientific superstition or should I say the better word would be the scientific hypnosis.

How both these theories Relativity and Quantum Mechanics are scientifically proved epicycles are discussed and explained in the next 8 books and what are their corresponding true physical phenomenon is finally presented in the last 10th book which presents a completely new physics, I name it the "Quassical Physics", without any postulates of relativity and without any postulate of quantum physics and without all the abstract mathematics, No Lorentz's transformation, no non-Euclidean geometries, no probabilities, no uncertainties, no matrices, no wave function, no complex algebra, no multi dimensions, no catastrophe, no Dirac notations, No Schrodinger's equation, no de Broglie matter wave, no quantum numbers, no paradoxes... and a lot more nos., of course, there are new postulates which are consistent with the Classical Physics, and some parts of true Quantum Physics, and entirely based on empirical knowledge and they will be open to be challenged on scientific and philosophical grounds. Seems unbelievable, because we have got used to live with abstract ideas without any physical understandings, because we have been taught that the world is like this only, because we believe the world is really like this.

The postulates of Quassical physics are conceptually simple, intuitive, deterministic, non-mathematical, observer independent, measurement independent, which do not produce infinities and paradoxes, obvious and easy to understand to everybody even to non-professional physicists. Once they are out, it will be very hard to neglect them on scientific and philosophical grounds.

This is a new revolution in physics, this is inevitable, this cannot be stopped, sooner or later these ideas / the new physics without relativity and quantum physics will have to be accepted, because, the new physics explains the physical world in a clean, consistent, scientific and non-mathematical physical intuitive way.

This is as if we have undone the physics of past whole century and have reached where the giants of true physics, to name a few, Copernicus, Kepler, Galileo, Newton have driven and left us, and taken a new path in continuation. This new theory will fill the gap of the century's lost progress and will open a new window of whole new physics, and I claim, it will save huge amount of effort and resources from the dis-oriented research in physics in future.

The Quassical physics makes the physics simple and intuitive, it removes all the mathematical complexities, but the side effect is, it makes the physics less interesting no time travel, no space-time warp, no parallel universe, no observers, no twins and no Bob and Alice. Pride should not be taken in complexity of solution of any problem, in true science, particularly in Physics. The aim should be to provide the explanation or solution in the simplest form so that majority of the people can understand it, professional or non-professional, to understand the Nature is not an exclusive right of the experts and professionals. The understanding may come from the non-professionals too, history has enough examples, but, whatever is the source of understanding, it must be tested on true scientific principles and on correct philosophy for validation.

15
VERY BRIEF HISTORY OF DEVELOPMENT OF THE FUNDAMENTAL THEORIES OF MODERN PHYSICS WITH PRESENT ACCEPTED VIEWS

To have an idea about where and how the problem originated, it is necessary to go back to the history. As mentioned many times earlier, There are two main theories in Modern Physics, one is General theory of Relativity which includes the Special theory of Relativity also and the other is theory of Quantum Mechanics which is the basis of other quantum Physical theories like QFT, QED, QCD, Standard Model etc., Let us explore some of them:

It is generally accepted that Relativity is applicable in macroscopic world and Quantum Mechanics is applicable in microscopic world. If both the theories are correct then it implies that somehow, the Nature has made two entirely different theories to explain the single Natural world, and the most astonishing thing about these theories is that, the boundary between the two worlds, macroscopic world and the microscopic world is not well known, and another most astonishing thing is that on theoretical grounds both the theories are absolutely incompatible and contradictory to each other.

There was another set of theories called the Classical theory of Physics, which is the grand old theory, the present status of the Classical Physics is a failed theory because it failed to explain certain experimental observations, but it is still useful in some areas of physics and seems to work correctly, for example one of the law of the Classical Physics, the Newton's law of gravitation works perfectly well within the whole solar system, it is practically enough to plan the trajectory of any space craft or any satellite within the solar system, even for all the planets of the solar system it works perfectly well practically. It is accepted that in certain conditions General Relativity produces the same results as the classical theory similarly in certain conditions the Quantum Mechanics also produces the same results as Classical Physics. Therefore, the Classical Physics is given the status of an approximate theory.

The accepted view of General relativity is that, it is mainly the theory of gravitation; it tells what the cause of gravitation is which Newton could not find out. Newton discovered the law of gravitation which is mainly based on Kepler's understandings and empirical findings of laws of planetary motions which are themselves are based on empirical findings of Tycho Brahe's precise measurements of planetary motions, therefore, Newton's law of gravitation is absolutely empirical, a result of direct communication with the Nature, direct communication with the physical world, the physical reality, but he could not provide the reason how gravity acts at distances because visibly there is no connectivity between the Sun and the planets, still the gravity of the Sun controls the movements of all the planets. Newton accepted openly that he does not have the answer to this question and left it for the posterity to answer. This was one of the great mysteries of the Classical Physics.

15.1   SPECIAL THEORY OF RELATIVITY

Another great mystery of Classical Physics was, what is light made of and how light moves in vacuum, Newton proposed that light is made of particles and explained some properties of light but failed to explain the refraction of light, which Christian Huygens was able to explain by assuming light not as particles but as waves, therefore, Newton's theory of light as particle was rejected and Huygens theory of light as waves was accepted on scientific basis, as it explained the observations of the physical world. Since light was accepted as a wave, it was assumed that there exists some sort of ubiquitous medium pervading the whole Universe, through which light propagates, then lot of efforts were made to prove the existence of this medium which is called the Aether, one such serious attempt was made by Michelson and Morley by using light interferometer, which is famously called the Michelson Morley experiment. This experiment was designed to prove the existence of the Aether by using the simple classical concept of relative velocity, the

Book 1: Why A New Revolution in Physics is Essential and why it is Inevitable

Earth moves at about 30 km/s in its orbit around the sun, it was supposed that the Aether around the sun is stationary and the Earth is moving in its orbit relative to this Aether with its orbital velocity of 30 km/s, the experiment was designed to detect the effect of this relative velocity between the stationary Aether and the Earth by using the light beams, assuming that light also follows the simple classical laws of relative velocities, but for the surprise of everyone, the result was unexpected, it produced null result as if either the Earth is not moving or the light is not following the simple classical law of relative velocity. Both the explanations seem unacceptable to explain the null results of the experiment. Then came Einstein, he was aware of the null result and also aware about the problems in classical electrodynamics which were also because of relative velocities, originating from Maxwell's equations of electromagnetism, which established that light and electromagnetic wave are the same thing. After some initial struggle Einstein decided in favor of the later one, light/electromagnetic wave does not follow the simple classical law of relative velocity. Simple. In my view, the question proposed by the Michelson Morley experiment and the classical electrodynamics were not understood correctly, Einstein assumed that the experimental result provides only two options out of which one needs to be selected and he selected the later one, but Einstein overlooked the third option – none of the above. Anyway, Einstein selected one of the options but it was impossible to reconcile with the well-established ideas of the Grand Old Classical Physics. It required the destruction of the whole Classical physics, I am not exaggerating, rather his destruction of the Classical Physics was extremely underestimated, I am surprised that how this extremely important change was not even discussed and talked openly? I am still surprised why the destruction of the Classical Physics was not resisted at all? Why the physicists at that time did not resist this non-scientific destruction?

    Now, let me explain what I mean by the destruction of the Classical physics. It was not that Einstein choose one of the two options visible to him and solved some problem and the life remained as good as it used to be earlier. Absolutely not. With the selection of one of the option, and to explain the null results of the Michelson Morley experiment based on the selected option that light do not follow the simple classical rules of relative velocity, Einstein had to modify and redefine the very fundamental concepts of the Classical Physics – space, time, simultaneity, length, mass and energy, one more very destructive change he introduced - every physical entity became velocity and reference frame / observer dependent. Einstein's move was very efficient, conceptually he simply rejected that light obeys the laws of Classical Physics and converted this concept into a mathematical theory by adopting the Lorentz's transformation, this way he invented (not discovered, in the sense Newton discovered the law of gravitation) a new theory – the Special theory of Relativity, which he presented in 1905. The modification and redefinition of the above mentioned fundamental concepts of the classical physics are only the interpretation of the mathematical theory based on the Lorentz's transformation. In my view, the whole hell broke loose in the name of new theory of the physics and its interpretation, many new fancy terms came into existence, time as a dimension, 4-dimensional space-time, flat space-time, mass swelling, time dilation, length contraction, the light cone, space like events, time like events, light like events etc. every such terms provided only mathematical knowledge with complete dearth of physical understanding, there was no guarantee that these new mathematical knowledge has anything to do with the understanding of physical reality. It must be noted that there may not be any relation between knowledge and understanding. It is impossible to make the connection between the two, that's why many new paradoxes came into existence – the twin paradox, the train and tunnel, the rivet and bug paradox etc., surprisingly, the mainstream physicists accepted these paradoxes, every attempts are made to explain these paradoxes but they couldn't succeed a bit, the fact remains that they are still correctly called paradoxes, literary contradicting themselves. One would be surprised to see that how the mainstream physicists made every effort to clear the paradoxes with every possible non-scientific means and ideas. Many believe the paradoxes have been solved and there remains no more paradoxes related to special relativity, in my view, this is a defeated blind faith. The fact is, not a single paradox has ever been solved in a true scientific sense. It is a very strange situation even at present, the theory and its proof of failure – the paradoxes, co-exist simultaneously. The mainstream physicists learned to live in this paradoxical situation, simply by denying and ignoring the proof of failure of the theory to move forward. The question arises – is it scientific? Is it physics? Is it the correct scientific way of dealing with any fundamental theory? Another very important question arises here – with the destruction of whole Classical Physics what did the Special theory of Relativity achieve? The astonishing fact is - absolutely nothing. Special theory of relativity only claims to explain the null result of Michelson Morley experiment and the conductor and magnet problem of Classical electromagnetics and nothing else from the whole lot of problem of physics. Doesn't it seem costly? It claimed to explain only the results of two experiments at the cost of a whole established Classical Physics! And the by-products of special relativity are much worse and in great many numbers than the two experiments it claimed to explain. An element of surprise is still left, the reasoning by which Einstein arrives at the special relativistic formula in1905 paper is itself based on non-scientific wrong reasoning, this is discussed in details in my another book of this series "The Collapse of Special theory of Relativity". For me, purely for personal reasons, there is still a great mystery left – why the mainstream physicists accepted and validated Special theory of Relativity in spite of its extremely limited scope of usefulness along with tons of paradoxes at the exorbitant cost of loss of the whole Classical Physics?

15.2 GENERAL THEORY OF RELATIVITY

The next forward move or progress (I am not using the term scientific progress) of physics came again from Einstein, when he invented the General theory of Relativity and presented it in 1916. The question arises – why Einstein chose only one problem to address and include it in special theory of relativity, while there was and still is a lot of unsolved basic problems like the cause of inertia, electric field, magnetic field, heat, momentum, potential energy etc. Anyway, Einstein chose only gravity to include in Special theory of Relativity to make it a fundamental theory of physics. Newton accepted that he does not know the cause of gravity, after that nobody seriously attempted to present any explanation of gravity, Einstein took the challenge and used something the Greeks were obsessed with - the Geometry. Einstein used geometry to explain the cause of gravity, geometry of what? Geometry of space-time! But how space-time and geometry can be connected? How can one understand this theory? Well, they are connected mathematically; using very complex mathematics and the theory can only be understood mathematically! The mathematical form of the General theory of Relativity is not created by Einstein alone, many prominent mathematicians have made their contributions in its development like, David Hilbert, Grossmann, Besso, Abraham, Nordstrom etc. In my view, this is the second invasion and first large scale daylight obvious invasion of mathematics on physics, the first invasion was in the form of Lorentz's transformation in Special theory of Relativity, an innocuous looking covert invasion. The fact is, the General theory of Relativity is so complex that no physicists understands the mathematics involves and no mathematician understands the physics involved, there is a famous remark of Einstein in this regard which I have already mentioned earlier but it worth mentioning again:

> "Since the mathematicians have invaded the theory of relativity, I do not understand it myself anymore."
> -Albert Einstein.

And, there is another famous anecdote which I have already mentioned above in this book, that there are only two persons who understood the General theory of Relativity, its main creator and it's claimed experimentally verifier. Later the creator had already confessed that he himself does not understand the General theory of Relativity, should we assume the verifier understood the theory? Going a little deep, it comes out that the verifier Arthur Eddington, was highly impressed from Einstein, and the details of its first confirmation experiment of the prediction of the General theory of Relativity, the solar eclipse experiment of 1919, was literally under clouds, there can be other simpler explanation of this experimental result. Then who is there to understand the General theory of Relativity? I understand that these comments are made in lighter note, but sometimes they contain the truth and only the truth which can't be told formally, this, in my view, is one such perfect example, it becomes more obvious retrospectively from today's status of physics.

But, how conceptually Einstein used the geometric characteristics to explain the cause of gravitation? Einstein claimed he encountered the happiest thought of his life; a freely falling person will not experience his weight, and remained besides himself on joy for many days. After this realization he invented the equivalence principle on which the whole General theory of Relativity is based conceptually. Then what is the equivalence principle? We need to go back to classical physics to understand this concept. In Classical Physics it was one of the biggest mystery that why the inertial mass and the gravitational mass are equal? As the anecdote goes, Galileo performed the famous experiment of dropping two balls of different density materials from the tower of Pisa to experimentally prove the equivalence of inertial and gravitational mass, whether the experiment was actually performed or not, but Galileo and his successors were well aware about this mystery of equivalence of inertial and gravitational mass, no understanding could be attained and no explanation could be provided. In Classical Physics, the inertial mass and the gravitational mass are two absolutely different physical properties of a mass, conceptually absolutely independent from each other, but experimentally it has been found exactly the same, this is a big unsolved mystery of the Classical Physics. It remained a mystery, and in my view, is still a mystery today for modern physics. What Einstein did was that, instead of solving the mystery and providing the scientific explanation, he assumed the mystery itself as a law of nature. He assumed that inertial mass and the gravitational mass are same, based on this assumption he invented a thought experiment and claimed that it is not possible to distinguish in any way between an accelerated system in the absence of any gravitational field and a system at rest in a gravitational field, and then like Special theory of Relativity, he converted this conceptual assumption into a mathematical theory by taking help of many mathematicians. The outcome is the General theory of Relativity, a highly complex mathematical theory. It is claimed that this is the theory of gravitation which is correct in all circumstances and it produces the same results as the Newton's theory of gravitation at small mass and low density conditions. It is also claimed that the Newton's theory of gravitation fails at large mass and high density conditions. Therefore, General theory of Relativity is the general theory of gravitation, applicable in all conditions and Newton's theory is claimed just an approximation which works well only within a limited conditions.

But how exactly, the general theory of relativity explains the cause of gravitation? It all comes from the interpretation of the mathematical equation of General theory of Relativity. It should be noted that the initial concepts on which the mathematical theory is based doesn't contain anything about the origin of gravitation. The strange thing is that, no physicist understands the mathematical equation of general relativity then how this interpretation is made that the equation of General theory of Relativity explains the origin of gravitation? Now, how I can claim that no physicist understands the mathematical equations of General theory of Relativity, which is called the field equations of general relativity, because I have already proved the field equation of general relativity mathematically wrong, which is given in details in my book of this series – "The Collapse of the General Theory of Relativity".

A New Revolution in Physics

Let us see, how the field equation of general relativity are interpreted to explain the cause of gravitation conceptually? We need to go back to Special theory of Relativity to start with, the Special theory of Relativity is the theory which doesn't take gravitation into account, it works in absence of gravitation, it's strange, but it means that Special theory of Relativity is not applicable to the real world! Because, gravitation is present almost everywhere in the whole Universe. Then what is it good for? Anyway, because the special theory of relativity is derived without taking gravity into account, the space-time of the Special Relativity is considered a flat space-time. Now what is space-time? This is out of the world concept of Special Relativity, nobody understands but nobody complains because it is deeply related to the Emperor's new clothes. It looks dead simple but is really killing, certainly killed the whole Classical Physics. Classical physics believes in 3 dimensional space which is absolutely independent of time, which was also Newton's view. Special Relativity combined the 3 dimensions of space and time and created 4-dimensional space-time, it is as simple as this. No, absolutely not, this step changed the whole Universe of physics, in my view, this is the biggest non-scientific conceptual attack on the Classical Physics, looks so simple but extremely lethal, the Classical Physics could not survived this attack. It is not possible to discuss at this place in details but is discussed in my other books of this series. So, this is the space-time, a 4-dimensional space-time Special Relativity has created in a single step, as mentioned above. This 4-dimensional space-time is considered flat in Special Relativity, but, in General Relativity this 4 dimensional space-time becomes curved and as per the interpretations of the field equations of the General Relativity, this curvature of 4-dimensional space-time is the cause of gravitation! Now, the question is, why does the space-time become curved? The answer is because of presence of mass. The presence of mass changes the 4-dimensional flat space-time nearby it to a 4-dimensional curved space-time, resulting appearance of gravitation, more the concentration of mass more is the curvature and therefore stronger the gravitation around the mass, the field equations of General Relativity doesn't see any limit, it predicts infinite density of mass which creates infinite curvature of space-time leading to infinite gravitation and thus creates another out of the world object, the very famous – the Black hole. Which has been found and its existence has been confirmed and accepted by the mainstream physicists by validating the recently first ever direct photograph taken by the event horizon telescope, if their claims are believed. Just imagine, what would happen to this photograph and other experimental confirmation of General Relativity like the detection of gravitational waves and the measurement of frame dragging if the field equations of the General theory of Relativity are proved wrong?

15.3 QUANTUM MECHANICS

Now let us move to a brief history of development of quantum mechanics / quantum physics.

The concept of quantum is all started with Max Planck's introduction of "Planck's constant" to create the mathematical model of the empirical black body radiation curves, he introduced this constant as a formal mathematical convenience to arrive at some mathematical equation which fits with the empirical data, he had absolutely no physical understanding of how his constant worked perfectly on empirical data, therefore I would like to say that Planck invented the "mathematical quantization". Let us see the contrast with relativity, in relativity its mathematics is not understood but the interpretation of this not understood mathematics claimed how the mathematics explains the physical reality is perfectly understood, while in Planck's case, the mathematics was completely understood but how the mathematics explains the physical reality was not understood. Planck's step was reflecting the characteristics of the physical reality, perfectly matching with empirical data without clouds and any large uncertainty, even though the mathematical steps taken by him were not understood by himself and others at that time, and Planck openly admitted it also. While Einstein's steps were claimed to be completely understood that how the complex mathematics of the field equations are explaining the cause of gravitation, but the basic difference is, in Einstein's case there is no absolutely clear match with the empirical data like in the case of Planck. Planck's claim was absolutely empirical still it was unbelievable to him, while Einstein's claim was absolutely non-empirical, based on interpretation of arbitrary mathematical equations which nobody understood, and he was sure of the correctness of the physical interpretations of these equations, the fact is, the mathematics the physics and all the claims of experimental confirmations of General Relativity are under heavy clouds. In my view, the field equations only explain the appearances and have no connections with the physical reality. It becomes important to quote Planck and Einstein in this regard:

> "It was 'an act of despair'... 'a purely formal assumption' "
> -Max Planck on introduction of "Plank's constant"

> "Since the mathematicians have invaded the theory of relativity, I do not understand it myself anymore."
> -Albert Einstein on General theory of Relativity.

> "...then I would be sorry for the God."
> -Albert Einstein on experimental confirmation of General theory of Relativity.

The step taken by Planck to explain the physical reality - the mathematical quantization whose physical counterpart was the quantization of energy, is considered the failure of the Classical Physics, because presumably quantization of energy is an entirely new idea not supported by the Classical Physics and the attempt to explain the black body radiation by Classical Physicists ended in ultraviolet catastrophe. In my view, this is not correct, which I have explained in details in my another book of this series – "The Resurrection of the Classical Physics".

But, it was Einstein's truly great scientific achievement, unlike Relativity, which in my view is non-scientific, that he explained another mystery of physics, the photo electric effect, for which he received his Nobel Prize. He used Planck's quantization of energy principle and claimed that light or electromagnetic energy exists in quantized form not during only the exchange as per Planck's view, but always, the credit goes to Einstein to convert the poorly understood concept of mathematical quantization of Planck into well understood actual quantization of physical reality, the energy of electromagnetic waves are quantized, how? is another story, which is still not understood clearly and that's why it is one of the main part of the New revolution of Physics which I have discussed in complete details in my most important book of the series – "The Concepts, The Dawn of the New Physics – the Quassical Physics".

It was also Einstein who first experimentally proved the existence of atoms by using and analyzing the Brownian motions, this is also another form of quantization, the quantization of mass. In my view, this is also very important achievement but somehow it could not gather the center stage it deserved. With the concept of atom, discovery of electron by J. J. Thomson and the discovery of Nucleus by Ernest Rutherford came the next big mystery of all time, the stability of atom. In my view, this is still not properly understood; therefore, this makes another major part of my most important book – "The Concepts, The Dawn of the New Physics– the Quassical Physics".

So far, the quantization is reasonably understood. It started with Plank's mathematical concept of quantization which Einstein converted into realistic physical concept of quantization and it successfully explained the physical reality of characteristics of the black body radiation and the photo electric effect.

The real mystery of the quantum physics started with the attempts to explain the stability of atom and to explain the characteristics of atom – the line spectrum. It is claimed that the Classical Physics failed to provide any scientific explanation, therefore, the whole Quantum Mechanics was requires to be invented to explain only these two mysteries.

Let us see, what exactly the mystery is and why it is claimed that the Classical Physics failed to explain it. J. J. Thomson has shown that an electron is negatively charged particle and, Rutherford has shown that atom is hollow and the nucleus is positively charged, and it was a well-known fact that an atom is electrically neutral, therefore, electrons must be outside the nucleus. It was also a well-known fact of the Classical Physics that opposite charges attract each other. So the mystery is, how a positively charged nucleus and the negatively charged electrons can form a stable atom, why not the electrons which are outside the nucleus get attracted by the nucleus and fall into nucleus? Classical Physics initially tried to explain with the model of solar system, in which the gravitational attractive force between the Sun and the planets is balanced by the centrifugal force generated by the orbital velocity of the planet around the Sun, proper balance of both the forces makes the stable orbit of the planet and the planet keeps revolving around the Sun. In the case of atom also, if the electrons are considered revolving around the nucleus then the centrifugal force created because of the orbital velocity can balance the electrostatic attractive force between the nucleus and the electron, both the forces will balance making a stable orbit of the electron around the nucleus, but there is a problem, the classical electrodynamics says that any accelerating charge radiates energy, therefore, a revolving electron will radiate its energy as it is under constant acceleration because of continuous change of direction of velocity. An electron continuously radiating its energy will lose its kinetic energy, therefore, it will not be able to maintain its orbital velocity and will fail to generate enough centrifugal force to counter the electrostatic force and will start falling towards the nucleus, finally all its kinetic energy will be radiated out and the electron will fall into the nucleus, making the atom unstable, but we know atoms exist, the electrons do not fall into the nucleus, what holds the electrons into stable orbits is the greatest mystery.

The solution of the mystery started with Niels Bohr; to explain the stability of atom he postulated "stationary orbits". When electrons move in these orbits then they do not radiate any energy and therefore its orbital velocity remains constant which maintains the centrifugal force created and balances the electrostatic attractive force and the electrons keep moving in the stable stationary orbits. Bohr proposed solution of the other mystery also, the mystery of atomic line spectrum, Bohr postulated that only certain stationary orbits are allowed with certain energies, not any orbit with any energy is allowed contrary to the Classical Physics, and he further proposed that light or electromagnetic wave is emitted when an electron moves from higher energy orbit to lower energy orbit which is seen as line in an atomic spectra. In short, this is called the Bohr model, it explained nicely the single electron hydrogen atom and other ions and atoms having only single electron, but this Bohr model did not work for any other atom which has two or more electrons, something had to be modified. It is worth noting here that the assumptions Bohr made are absolutely arbitrary and contrary to the known laws the Classical Physics of the time, and there was no scientific basis for these assumption. Therefore, this step taken by Bohr cannot be called a scientific step, he did not give the explanation of the problem, on the contrary, he converted the problem, required to be explained, into new laws of physics! Exactly as Einstein did in the Special theory of Relativity in second postulate about speed of light, and in the General theory of Relativity about the equivalence principle. He created a new non-scientific culture which had more staunched followers who were going to surprise Einstein himself.

Now came Louis de Broglie, he was doing Ph. D. thesis on dual nature of light, which claims that light behaves as particles in some experiments and as waves in some other experiments, this is called the dual nature of light. The dual nature of light is accepted as experimental fact, which was known to the Classical Physics clearly. The unconventional claim made by de Broglie was that, the electron which is regarded as a particle with certain mass, and other sub atomic particles, also has dual nature, he derived a mathematical formula for the wavelength of this "matter wave" using mass energy equivalence principle which is considered an outcome of Special theory of Relativity. then G. P. Thomson, the son of J. J. Thomson, showed that electrons diffracts from some crystals just like waves, showing wave nature of electron experimentally, later Davison and Germer also performed an experiment which showed wave nature of electron, they performed double slit experiment and used electrons in place of light and they were able to get the interference fringes which provided confirmation of wave phenomenon. These experiments supported the claims of de Broglie and the dual nature of sub atomic particles was established. Louis de Broglie proposed the dual nature is not limited to only sub atomic particles, it should be valid for other matter also, in my view, de Broglie's this proposal set the stage for quantum magic to be performed.

Confident with experimental support of his ideas, de Broglie proposed that, Bohr's arbitrary postulates of stationary orbits can be avoided if electron can be considered as wave, and he proposed orbiting electron are not a particle but a closed circular standing wave. This hypothesis claimed that an orbiting electron is not a point mass which occupies a tiny space at a specific position at some specific time, rather it spreads out and fills the whole orbit and thus it escapes the pull of the nucleus and therefore, Bohr's baseless arbitrary postulate of stationary orbits are not required, and thus the electrons can be prevented from falling into the nucleus. He proposed different modes of vibration of the stationary wave can explain the orbits with specific quantized energies. Thus the two great mysteries are solved by using de Broglie's hypothesis of matter waves without any arbitrary assumption of Bohr and the stability of the atom is saved. Prima facie, this looks great that the job is done, but it brings many problems which is difficult to understand, like how the charge of the electron should be considered, can it be considered concentrated point like or can it be distributed in the entire orbit, but the charge of an electron is defined fundamental unit of electric charge which means, it is the minimum amount of quantized charge available, it is also claimed that no experiments have so far detected a fraction of charge of an electron, at this point quarks can be pointed out but they do not exist isolated as per the theory, so they don't fit to be mentioned here, there are several other issues with this view which is discussed in details in my another book of this series – The

Collapse of the Quantum Physics.

Empirically it was known that atom has a very specific distribution of electrons around the nucleus, then how Bohr met these requirements? He did the reverse engineering and postulated one more law, quantization of angular momentum, this helped Bohr to distribute the electrons as per the empirical findings, It should be noted that, Bohr postulated three laws, stationary orbits for electrons, only certain stationary orbits with certain specific energies are allowed and quantization of angular momentum of electron. These three laws can meet the requirement of empirical findings of the distribution of electrons around the nucleus, but there was a big problem, they can only account for half the number of electrons, another law was required, the fourth required law came from the experimental results of Stern Gerlach experiment, it created another out of the world concept – spin. It was established that electrons have a strange property which can only assume two values, up or down, this quantum spin is absolutely different than the normal classical concept of spin which is nothing but rotation about its own axis. It is claimed that spin is the property of electron not of an atom but it affects the distribution of electrons in an atom. Bohr adopted this as fourth law and the number of electrons matched perfectly, each of these four laws is assigned a quantum number, therefore. Thus Bohr's model of atom is based on 4 postulates whose origin are unknown and violates the laws of the Classical Physics; therefore, the postulates are based on empirical findings. All these, only for hydrogen atom, a one electron atom. Bohr's model can explain the atomic structure of only single electron atoms or ions.

Meanwhile, Warner Heisenberg has created another pillar of quantum mechanics, he has, in my view, "invented" the uncertainty principle, according to which, there is a theoretical minimum limit of attainable knowledge of the quantum world, and no knowledge can be attained below the minimum limit of a quantum system or of the quantum world. The important point to be noted is that, the uncertainty principle claims the minimum limit is theoretical, it is not technological. According to the uncertainty principle, there are paired quantities in the quantum world and both the quantities cannot be measured simultaneously with unlimited precision, if one of the quantity of the pair is measured with more precision then uncertainty of the measurement of the other quantity of the pair becomes high so that the product of the uncertainties of both the quantities remains constant. In my view, this is also an out of the world hypothesis. It would be appropriate to mention here that Heisenberg was highly impressed with the non-commutative properties of matrix multiplication, he considered it as one of the fundamental law of the Nature, he discussed this with Bohr but Bohr was not impressed, then he invented the thought experiment -Heisenberg's microscope to build the idea and finally proposed the uncertainty principle whose mathematical form contains the commutators, the most devastating step taken by Heisenberg, in my view, is his claim that the uncertainty principle is the fundamental law of Nature of the quantum world, and they, including Bohr and others frontiers of the quantum mechanics, made the uncertainty principle the foundation pillar of the Quantum Mechanics, a lot of Quantum Mechanics and Quantum Physics are based on this principle, It should be noted here that although the uncertainty principle is made the foundational pillar of Quantum Physics but the required conceptual and philosophical clarity is completely missing. In my view, it doesn't qualify as a fundamental principle which I have discussed in details in my book of this series – "The Collapse of the Quantum Physics".

Came Schrodinger, and with him another level of quantum surprises. He proposed that an electron in the atom behaves as a 3 dimensional stationary waves and fills regions of space around the nucleus. By using the de Broglie's concept of matter wave and Heisenberg's uncertainty principle, Schrodinger created a wave equation, a partial differential equation which used complex number. Now, in my view, this step of Schrodinger is a non-scientific step and I have discussed it in another book of this series – "The Collapse of Relation between Physics and Mathematics"

But, the problems is, the exact solution of the Schrodinger's equations are extremely difficult, almost impossible, It has only been solved exactly for hydrogen atom, for other atoms either approximate methods or numerical methods are required to be used which require huge computational powers, even modern supercomputers fails to meet the demand. Now the question is - should this be called a scientific theory? Which can't even be solved for more than one or two cases and almost all the cases remain unsolved, why shouldn't it be called a calculated fluke which works for one or two cases? It is claimed that the quantum numbers naturally appears out of the Schrodinger's equation, and this is regarded the success of the Schrodinger's equation, this means, somehow the Schrodinger's equation is deeply connected to the reality of the quantum world, it reflects the quantum reality. The solution of the Schrodinger's equation produced regions of space around the nucleus, the shapes and sizes are linked with the naturally appearing quantum numbers, these are called the orbitals, but, one problem was still haunting, in what form the electrons exist around the nucleus in an atom? Are these regions of space, the orbitals, the electron clouds of a single electron somehow spread and filled the whole orbital space? Or are they only the permissible regions allowed for a single particle electron to move? There are several unanswered questions. Although, the solution of the Schrodinger's equation explained the structure of one electron hydrogen atom, But, the question was still open and uncomfortable – in what form the electrons exist around the nucleus in an atom, what exactly are the orbitals produced by the solution of the Schrodinger's equation?

Then came Born for rescue, Born's proposal was mind blowing, he proposed that the orbitals are the probabilities of finding an electron around the nucleus! In my view, this is another non-scientific step in the development of quantum mechanics. But the fact is, this is the final and current accepted version of the quantum mechanics.

Going through this very brief history of quantum mechanics, which covered the important milestones and important steps, does one feel the foundation of quantum mechanics is as solid as the foundation of the so called failed Classical Physics? The empirical laws of Classical Physics are frozen although their origin are not completely understood, in my view, we won't see any change in Newton's law of Gravitation, Coulomb's law of electrostatics force, law of conservation of linear momentum, law of conservation of angular momentum, law of conservation of energy etc. practically they are frozen. Quantum mechanics doesn't even know in what form an electron exists in the atom (outside the nucleus), it doesn't even know what the basic quanta of energy, the photon is – a wave or a particle. To summarize, the Quantum mechanics is the drama played by photon, electron, proton, and neutron but it doesn't know what exactly these characters are and what part they are playing, the whole plot is in the grips of uncertainty. Similarly, General Relativity which claims to change the Newton's law of gravitation don't seem to have a solid foundation needed to perform the claimed task, it cannot explain conceptually how a 4 dimensional space-time curves into the 4 dimensions itself and creates the gravitation? In my view, it seems it doesn't even have a foundation at all, what is the basis of my claim is discussed in details in my book – "The Collapse of the General theory of Relativity".

16
WHY DIFFERENT FUNDAMENTAL THEORIES OF THE MODERN PHYSICS WERE REQUIRED TO BE CREATED?

Let us summaries in concise way why several fundamental theories are required to be created, it will be surprising that each theory was developed to address only one or two problems, but later on they were used to explain many different phenomenon and thus claimed they are general fundamental theories which can explain many empirical findings or problems or mysteries of the real physical world.

Let us see what were the main problems to solve them these theories were created

	Theory	Problems which lead the creation of the theory
A	Special theory of relativity (Einstein)	To explain the null result of Michael Morley experiment. Mainly one problem, the others are afterthought or by product
B	General theory of relativity (Einstein)	The cause of gravitation. Mainly one problem, the others are afterthought or by product
C	Quantum Physics (Pre) (before the postulates of Bohr) (Max Planck)	The Blackbody radiation characteristics, Mathematical quantisation of energy, no physical Understanding

	(Einstein)	Photo electric effect, physical quantisation of energy physical understanding.
D	Quantum Mechanics (Bohr, Heisenberg, Pauli, de Broglie, Schrodinger and Born)	The Stability of atom and atomic line spectra.

:

17
WHAT THESE FUNDAMENTAL THEORIES OF MODERN PHYSICS CLAIM TO EXPLAIN?

As explained above these theories are created to explain some specific problems/ phenomenon, but once created, they are claimed to explain several other phenomenon, following table gives the claim of the various theories, the initial entries in italics are the ones for which the theory was created and the other ones are the extension of the application of the theories:.

	Theory	Claims to Explain	
A	Special theory of relativity	*The null result of Michael Morley experiment.*	
		The conductor and magnet problem of classical electrodynamics (from Maxwell's Equations).	
		Nature of Space	
		Nature of Time	
		Nature of simultaneity	
		Nature of Vacuum, it does not exist	

		Maximum possible speed	
		Mass energy equivalence	
B	General theory of relativity	*The cause of gravitation*	
		Perihelion shift of Mercury	
		Nature of Space,	
		Nature of Time	
		Motion of a body in a gravitation field, Geodesic,	
		Bending of light by gravity, gravitational lancing	
		Equivalence of inertial and gravitational mass (by making postulate)	
		Expansion of the Universe	
C	Quantum Physics (Pre) (before the postulates of Bohr) (Max Planck)	*The Blackbody radiation characteristics Mathematical quantisation of energy, no physical understanding*	
	(Einstein)	*Photo electric effect, physical quantisation of energy physical understanding.*	
D	Quantum Mechanics (Bohr, Heisenberg, Pauli, de Broglie, Schrodinger and Born)	*The Stability of atom*	
		Atomic line spectra	
		The quantum numbers	
		The Compton effect	

Book 1: Why A New Revolution in Physics is Essential and why it is Inevitable

		The Zeeman effect, splitting of atomic spectral lines	
		Electron diffraction by a crystal	
		The structure of empirical periodic table	
		The atomic and molecular bonds and therefore the whole chemistry	
		Laser	
		The interference of Young's double slit experiment	
		Characteristics of light through two polarizers	
		The Bohr's postulated quantum numbers	
		Diffraction of single slit experiment	

18
ARE THE CLAIMS OF THE FUNDAMENTAL THEORIES OF MODERN PHYSICS SCIENTIFICALLY CORRECT?

Following is the result of true scientific analysis, how this conclusion is reached is discussed in complete details in the other relevant books of this series some of them are already discussed in this book itself:

	Theory	Claims to Explain	Scientifically right or wrong?
A	Special theory of relativity	The null result of Michael Morley experiment	Wrong
		The conductor and magnet problem of classical electrodynamics (from Maxwell's Equations)	Right
		Nature of Space	Wrong
		Nature of Time	Wrong
		Nature of simultaneity	Wrong
		Nature of Vacuum, it does not exist	Wrong

Book 1: Why A New Revolution in Physics is Essential and why it is Inevitable

		Maximum possible speed	Wrong
		Mass energy equivalence	Right
B	General theory of relativity	*The cause of gravitation*	Wrong
		Perihelion shift of Mercury	Wrong
		Nature of Space,	Wrong
		Nature of Time	Wrong
		Motion of a body in a gravitation field, Geodesic	Wrong
		Bending of light by gravity, gravitational lancing	Wrong
		Equivalence of inertial and gravitational mass (by making postulate)	Wrong
		Expansion of the Universe	Wrong
C	Quantum Physics (Pre) (before the postulates of Bohr) (Max Planck)	*The Blackbody radiation characteristics Mathematical quantisation of energy, no physical understanding*	Partially Right
	(Einstein)	*Photo electric effect, physical quantisation of energy Physical understanding.*	Right
D	Quantum Mechanics (Bohr, Heisenberg, Pauli, de Broglie, Schrodinger and Born)	*The Stability of atom*	Wrong
		Atomic line spectra (only single electron atoms and	Partially Right

		ions)	
		The quantum numbers	Wrong
		The Compton effect	Wrong
		The Zeeman effect, splitting of atomic spectral lines	Right
		Electron diffraction by a crystal	Wrong
		The structure of empirical periodic table	Wrong
		The atomic and molecular bonds and therefore the whole chemistry	Wrong
		Laser	Right
		The interference of Young's double slit experiment	Wrong
		Diffraction of single slit experiment	Wrong
		Characteristics of light through two polarizers, Dirac notations.	Wrong

19
ARE THE PRESENT ACCEPTED FUNDAMENTAL THEORIES DEVELOPED ON SCIENTIFIC PRINCIPLES?

The answer is very surprising – NO.

As I have already mentioned in several occasion in this book also why my answer is negative, there are several critical steps in the development of these theories which are non-scientific but which influenced critically the direction and final form of the theories. Each theory, Special theory of Relativity, General theory of Relativity, Quantum Mechanics, Quantum Physics, and Standard Model is discussed with every critical non-scientific step in complete details in the corresponding books of this series.

20
THE CLASSICAL PHYSICS CONCEPTS WHICH THE MODERN PHYSICS HAS CHANGED

Following table summarizes the main concepts of the Classical Physics which has been changed by different theories of Modern Physics

	Theory	Classical Physics concepts	Classical Physics concepts changed by Modern Physics
A	Special theory of relativity	The Aether. Universal existence of Aether for propagation of light in the Vacuum	Aether does not exist
		Relative velocity obeys Galilean Transformation	Relative velocity obey Lorentz's Transformation
		The nature of Space, -independent of time. -Independent of	The nature of Space, -not independent of time. -Depends on

		relative velocity	relative velocity
		The nature of Time, -independent of space. -independent of relative velocity	The nature of Time, -Not independent of space. -Depends on relative velocity
		Nature of mass, -independent of relative velocity	Nature of mass, -Depends on relative velocity
		Space is 3 dimensional Euclidean	Replaced by 4 dimensional flat space-time
		Nature of Simultaneity, -Existence of universal "Now". -Independent of relative velocity	Nature of Simultaneity, -Rejects universal "Now". -Depends on relative velocity
		Time is not a dimension	Time is a dimension -Time is a fourth dimension. -It is equivalent as dimension of Space
		Space and Time are independent	Space and Time are not independent. -Space and Time can be added as space-time
		There is no maximum speed limit	There is a universal maximum speed limit. -Nothing can move faster than light

B	General theory of relativity	Newton's law of gravitation. -Universally applicable.	Replaced by Field equations of General Relativity. Claimed Universal applicability. Newton's law of gravitation is claimed only as an approximate law, which Works only in low mass condition Fails near large mass or high mass density conditions
		Gravity is a force.	Gravity is not a force. -Gravity is curved space-time
		Space is 3 dimensional Euclidean	3 dimensional space is replaced by 4 dimensional curved space-time
		Planets move in a curved orbit around the sun, under the influence of the Sun's gravitational field.	Planets do NOT move in a curved orbit around the sun, they move in a straight line along the geodesic in a 4-dimensional curved space-time wrapped by the Sun's mass
		The gravitational attraction force between the Sun and a Planet is keeping the planet	As per General theory of relativity there exists no gravitational force

		in the orbit against centrifugal force and prevents it from flying away.	and there exists no centrifugal force on an orbiting planet, the planets only follow the geometry of curved space-time around the Sun!!!
		An orbiting planet experiences centrifugal force.	There is no centrifugal force on an orbiting planet. -Centrifugal force is an illusion.
		Inertial mass and gravitational mass of a body are entirely different concepts, -Why they are equal is a big mystery.	Inertial mass and gravitational mass of a body are the same thing. Instead of solving the mystery and providing explanation, this is accepted as the basic postulate of General Relativity as Equivalence Principle
		Only mass creates gravitation.	Not only mass, but other forms of energy also create gravitation.
		Gravitation and time are independent.	Gravitation and time are NOT independent. -Gravitation affects pace of

			time called gravitational time dilation.
		An accelerating mass doesn't lose its energy.	An accelerating mass loses its energy in the form of Gravitational waves.
		A rotating mass does not affect the space around itself.	A rotating mass does affect the space-time around itself, it drags the space-time along with itself, -Frame dragging and Geodetic effects.
C	Quantum Physics (Pre) (before the postulates of Bohr) (Max Planck)	Energy is allowed to have any value; during exchange of energy every possible magnitude of energy is possible.	Energy is allowed to have any value, but, during exchange of energy every possible magnitude of energy is NOT allowed. Energy in multiple of a certain minimum constant chunk is allowed. -Mathematical quantisation of energy. -no physical understanding
	(Einstein)		Physical quantisation

Book 1: Why A New Revolution in Physics is Essential and why it is Inevitable

			of energy. -Physical understanding. Energy is quantised always.
D	Quantum Mechanics (Bohr, Heisenberg, Pauli)	There exists a real physical world independent of the existence of any living being, mind or measurements.	The quantum world does not exist. There is only quantum mechanical description.
		The aim of physics is to find the laws of the Nature.	The aim of physics is NOT to find the laws of the Nature; it's only what can be said about the Nature.
		The physical world is governed by the deterministic laws, causality rules.	The quantum world is governed by the probabilistic laws, uncertainty rules.
		Moon exists even when nobody is looking at it.	Moon does NOT exist when nobody is looking at it; it exists only when we look at it.
		Classical physics believes in a single seamless physical world from deep microscopic to deep macroscopic, single set of rules must prevail.	Modern Physics has different rules for the microscopic world and the macroscopic world. The biggest problem of the modern physics is that,

				the rules of both the worlds are incompatible to each other.
			A cat cannot be live and dead at the same time inside a closed box.	A cat can be both alive and dead at the same time inside a closed box. This is called Quantum Superposition, one of the very important pillars of the quantum mechanics!!! The Schrodinger's cat
			Act of measurement does not affect the outcome of any experiment (excluding simple observer's effect).	Act of measurement does AFFECT the outcome of any experiment (excluding simple observer's effect), -It collapses the wave function, it is also known as the measurement problem.
			Atom is unstable. An accelerating charge radiates electromagnetic radiation; an electron orbiting the nucleus will fall to nucleus. It is claimed that the	Atom is stable. Postulate of "Stationary orbits" prevents the fall of electron In a "Stationary orbit" accelerating charge does NOT

Book 1: Why A New Revolution in Physics is Essential and why it is Inevitable

		Classical Physics could not explain the stability of an atom.	radiates electromagnetic radiation within an atom.
		Any orbit of electron is possible.	Any orbit of electron is NOT possible. Postulate of quantised orbits, orbits with only certain energy is possible.
		An electron will move from one orbit to another orbit by continuous translation along a path, it exists continuously in the whole path.	An electron will move from one orbit to another orbit in discontinuous manner along a path, it does not exists in between for certain length of the path, it disappears from one orbit and suddenly appears at another orbit without passing in between, this is the postulate of the Quantum Jump!
	(de Broglie)	Dual nature is not possible for particles. Particle nature and wave nature is mutually exclusive	Dual nature is not only possible, but it is a law of nature for particles. It is called the wave-

		properties for all particles except light, (Photon). -Photon has dual nature, both wave nature and particle nature.	particle duality. -The corresponding wave of a particle is called the matter wave; the wavelength of a particle is governed by the momentum of the particle.
	(de Broglie)	Electron as a usual charged particle Electron is revolving around the nucleus as a charged particle, but the atom is unstable.	Electron as a 2 dimensional wave. Electron is not revolving as a charged particle; it exists as a 2 dimensional closed stationary wave filling the orbit.
	(Schrodinger)	Electron is revolving around the nucleus as a charged particle, but the atom is unstable.	Electron as a 3 dimensional wave. Electron is not revolving as a charged particle; it exists as a 3 dimensional closed stationary wave filling some special regions of space called the orbitals, predicted by the Schrodinger's wave equations.

Book 1: Why A New Revolution in Physics is Essential and why it is Inevitable

	(Born)	Electron is revolving around the nucleus as a charged particle, but the atom is unstable.	Electron is revolving around the nucleus as a charged particle, but the atom is STABLE. The orbitals are now interpreted as the probability of finding the electron around the nucleus.
		A particle moves and exists continuously in the whole path.	Electron moves in Quantum jump. An electron moves from one orbital to another orbital, or different parts of the same orbital which not not connected in Quantum jumps without passing in between!!!
	(Heisenberg)	There is no theoretical limit to acquire knowledge about anything, presently it is limited by the state of the technology.	There is a minimum theoretical limit to acquire knowledge about any quantum system. This is the Heisenberg's Uncertainty Principle, very important pillar of quantum mechanics.

21
THE NEW CONCEPTS WHICH THE MODERN PHYSICS INTRODUCED

Following table summarizes some new concepts introduced by different theories, of the Modern Physics the list is not exhaustive; some of the concepts may be common with the previous table:

	Theory	New concepts introduced
A	Special theory of relativity	Space is relative, depends on velocity, Length contraction
		Time is relative, depends on velocity Time dilation
		Mass is relative, depends on velocity Mass swelling, Relativistic mass
		Simultaneity is relative, depends on velocity No universal "Now"
		All inertial reference frames are equivalent, every observer's point of view on every inertial reference frames are equally valid

Book 1: Why A New Revolution in Physics is Essential and why it is Inevitable

		Nothing can go faster than light
		Mass energy equivalence
		Absoluteness of speed of light
		Proper time
		4-dimensional flat space-time
		Massless photons, massless particles
		Light cone, space like, time like and light like events.
		No Aether, Vacuum does not exist.
B	General theory of relativity	Equivalence principle Postulated as a law of the Nature
		4-dimensional curved space-time
		Light bends in a gravitational field gravitational lensing
		Gravity is curved space-time
		Mass creates curvature in 4-D space-time
		Curved space-time guides mass to move in a gravitational field
		A freely falling body follows straight line, the geodesic, in 4-d curved space-time
		Gravity is not a force
		Newton's law of gravitation is just an approximation, General relativity is the general law valid in every conditions, GR produces the same results as Newton's law for small masses
		Accelerating mass creates gravitational waves

		GR allows implosion which creates a Black hole and singularity.
		A spinning mass creates frame dragging of 4-D curved space-time.
		Gravity affects time and causes gravitational time dilation
		Gravity causes gravitational red shift
		Cosmological constant
C	Quantum Physics (Pre) (before the postulates of Bohr) (Max Planck)	The Blackbody radiation characteristics Mathematical quantisation of energy, no physical understanding
	(Einstein)	Photo electric effect, physical quantisation of energy Physical understanding.
D	Quantum Mechanics (Bohr, Heisenberg, Pauli, de Broglie, Schrodinger and Born)	Quantum numbers
		Principle quantum number
		Angular momentum quantum number
		Magnetic quantum number
		Spin quantum number
		Electron cloud
		H fine structure
		H hyperfine structure
		H Lamb shift
		Wavicle

Book 1: Why A New Revolution in Physics is Essential and why it is Inevitable

		Eigenvalue
		Eigenvector
		Hermition Operator
		Entanglement
		Superposition of states
		Wave function
		Collapse of wave functions
		Shut up and calculate
		Wave function Postulate Everything that we can know about a system is described by its wave function
		Operator postulate Every measurable physical quantity M is described by an operator M hat
		Local reality, local causality, locality
		Experimental results depend only on the local environment during the experiment
		Quantum superposition
		The Born interpretation: the wave function is the probability amplitude, the absolute square of the wave function is called the probability density And the probability density times a volume element in three dimension space is the probability.
		"Normalisation condition for the wave function"

		Position operator
		Momentum operator
		Energy operator
		Orthogonally theorem: Eigen functions of a Hermitian operator are orthogonal if they have different eigenvalues.
		Quantisation of the orbital angular momentum. mvr = nh
		Copenhagen interpretation
		Relative state (Multiple worlds, Decoherence)
		Transactional(Cramer) Physically real waves moving backwards in time
		Quantum information theory (Zero worlds): extension of classical information theory with complex numbers.
		The second derivative of a wave function is proportional to the kinetic energy operator
		More nodes of a wave function means higher energy
		The Orbitals Like an orbit of the Classical physics but it is a mathematical concept, the position of the electron in hydrogen atom is not at all as well defined as in a classical orbit.
		Momentum is gradient of position wave function
		Quantum Measurements

Book 1: Why A New Revolution in Physics is Essential and why it is Inevitable

		Observable
		Upness and downness, twoness
		Hybridisation
		Spin wave function
		Spin operator
		Spherical Harmonic wave function
		Heisenberg microscope
		Quantum objects are waves (as long as you don't look)
		Quantum revolution: transistor, laser, computer, GPS, MRI
		Sigma bond, pi bond
		Photons have spins, spin implies circular polarization, linear polarization
		Molecular wave functions
		Valance bond wave functions
		Observer effect
		Shielding
		Penetration
		Every possible wave function in existence can be thought of as a combination of an infinite number of functions that are each zero at all points except one!!
		Path integral formulation of Feynman
		Quantum eraser
		According to Quantum Mechanics photon exchanged between charged particles create force, thereby creating the

		illusion of electric and magnetic fields.
		"Normalisation condition for the wave function"
		Time frequency ambiguity
		Information is never lost
E	Quantum Physics(Post), QFT, QCD, QED (Dirac, Feynman)	Vacuum energy
		Particles are excited states of an underlying physical field, so these are called field quants
		QED has one electron field and one photon field,
		QCD has one field for each type of quark
		In condensed matter there is an atomic displacement field that gives rise to phonon particles
		The electron itself is just an excitation, a vibration
		Virtual particles
		A photon is an excitation. a vibration in an electromagnetic field
		Feynman's path integral method
F	Standard Model	Stability of nucleus - the strong force
		Emission of beta particles from the nucleus – the weak force
		Structure of proton and neutron
		Cause of mass of fundamental particle, the Higgs field

Book 1: Why A New Revolution in Physics is Essential and why it is Inevitable

		Quark, anti-quark
		Gluon
		Strong interaction
		Mesons
		Beta decay
		Time frequency ambiguity
		Each new rug of the ladder represents the existence of one additional particle
		Only 1% of the mass of a proton comes from its constituents quarks!

22

THE CONCEPTS OR PROBLEMS WHICH THE FUNDAMENTAL THEORIES OF MODERN PHYSICS DO NOT EXPLAIN BUT THE NEW QUASSICAL PHYSICS EXPLAINS

It is very surprising to realize the fact that how little these fundamental theories claim to explain and how many of them are scientifically correct? There are lots of phenomena these accepted fundamental theories do not explain, they don't have the general characteristics to address these phenomenon. It becomes reasonable to ask, should they be called fundamental theories at all? Let us compare how many other main concepts these fundamental theories of Modern Physics and the new physics –The Quassical Physics claim to explain:.

Concept	Theories of Modern Physics					New Physics
Concept	SR	GR	QM	QFT	SM	Quassical Physics
What is Space	No	No	No	No	No	YES
What is Time	No	No	No	No	No	YES
The cause of Inertia	No	No	No	No	No	YES
What is Mass	YES	YES	No	No	YES	YES
What is Charge	No	No	No	No	No	YES

Book 1: Why A New Revolution in Physics is Essential and why it is Inevitable

Concept	SR	GR	QM	QFT	SM	Quassical Physics
What is Velocity	No	No	No	No	No	YES
What is Acceleration	No	No	No	No	No	YES
Why acceleration needs energy	No	No	No	No	No	YES
What is an absolute reference frame	No	No	No	No	No	YES
The cause of Conservation of linear momentum (Empirical finding)	No	No	No	No	No	YES
The cause of Conservation of Angular momentum (Empirical finding)	No	No	No	No	No	YES
How a force works	No	No	No	No	No	YES
How action at a distance works (Newton's nightmare, Empirical finding)	No	No	YES	No	No	YES
The cause of equivalence of inertial and gravitational mass (Galileo's finding), empirical finding	No	No	No	No	No	YES
How velocity imparts KE to a mass	No	No	No	No	No	YES
What exactly PE in a field is	No	No	No	No	No	YES
What exactly is a field	No	No	No	YES	No	YES
What is gravitation field	No	YES	No	No	No	YES
What is electric field	No	No	YES	No	No	YES
What is magnetic field	No	No	YES	No	No	YES

Concept	SR	GR	QM	QFT	SM	Quassical Physics
How KE & PE converts to one another	No	No	No	No	No	YES
How gravitational field creates a force on a stationary mass (Empirical finding)	No	No	No	No	No	YES
The cause of inverse square law (Empirical Law)	No	No	No	No	No	YES
What exactly is a photon, a wave or a particle (the greatest mystery of Physics of all time, cause of origin of quantum mechanics)	No	No	No	No	No	YES
How a photon or an electromagnetic wave travels through the vacuum	No	No	No	No	No	YES
The reason of constancy of speed of light (Einstein's nightmare, Empirical finding, second postulate of special relativity)	No	No	No	No	No	YES
The physical interpretation of blackbody radiation characteristics (Planck's nightmare, empirical Law, mathematical solution and wrong physical interpretation)	No	No	No	No	No	YES

Book 1: Why A New Revolution in Physics is Essential and why it is Inevitable

Concept	SR	GR	QM	QFT	SM	Quassical Physics
The reason, what prevents the fall of electron to the nucleus (Bohr's nightmare, Empirical finding)	No	No	YES	No	No	YES
The cause of 3 quantum numbers (Bohr's nightmare, solved by converting them as postulates which created the whole quantum mechanics)	No	No	YES	No	No	YES
The cause of fourth quantum number, spin, the twoness of an electron (Empirical Law)	No	No	No	No	No	YES
What exactly is spin?	No	No	No	No	No	YES
The cause of the quantum jump (Schrodinger's nightmare, quantum magic)	No	No	No	No	No	YES
The cause of Pauli's exclusion principle (Empirical law)	No	No	No	No	No	YES
The cause of Afbua principle(Empirical Law)	No	No	No	No	No	YES
The cause of Hund's law(Empirical Law)	No	No	No	No	No	YES
The failure of quantum mechanics prediction why there is no 3d atoms after 2p atoms	No	No	No	No	No	YES

A New Revolution in Physics

Concept	SR	GR	QM	QFT	SM	Quassical Physics
The cause of structure of periodic table (Empirical finding)	No	No	YES	No	No	YES
The cause of x ray emission (empirical)	No	No	YES	No	No	YES
The cause of emission of energy when a bond is formed	No	No	YES	No	No	YES
The cause of octant rule	No	No	No	No	No	YES
The cause of hyper valent atoms	No	No	No	No	No	YES
The cause of hybridization	No	No	No	No	No	YES
The cause of strong force	No	No	No	No	No	YES
The cause of weak force	No	No	No	No	No	YES
The cause of gamma ray emission from nucleus	No	No	No	No	No	YES
What exactly is heat	No	No	No	No	No	YES
The cause of thermal expansion	No	No	No	No	No	YES
What exactly is temperature	No	No	No	No	No	YES
What is the structure of a photon	No	No	No	No	No	YES
What exactly is the wavelength of a photon	No	No	No	No	No	YES
How a photon carries momentum	No	No	No	No	No	YES

Book 1: Why A New Revolution in Physics is Essential and why it is Inevitable

Concept	SR	GR	QM	QFT	SM	Quassical Physics
How exactly an atom emits a photon (Schrodinger's nightmare)	No	No	No	No	No	YES
The cause of single slit interference	No	No	YES	No	No	YES
The cause of double slit interference	No	No	YES	No	No	YES
The cause of dispersion of light (Newton's Empirical finding)	No	No	No	No	No	YES

.... and many more, from the classical to present day phenomena.

It should be noted that, some of the phenomenon are explained by the modern physics and the new Physics, in majority of cases the explanations provided by the new Physics are entirely different.

.

23
WHAT ABOUT THE EXPERIMENTAL CONFIRMATIONS OF THE FUNDAMENTAL THEORIES?

This is also very surprising and seems unbelievable that all the experiment confirmations of all these fundamental theories have problems. many major experiments have been discussed in details in the corresponding books, scientific scrutiny are performed for the claimed results and working of the experimental set up also, the results are very surprising. Interpretation of the experimental result plays major role in many cases. The bottom line is, the claimed experimental confirmation of predictions of these theories are subjected to many uncertainties and non-scientific subjective interpretations, therefore, the words "experimental confirmations" have almost lost their true meaning..

24
WHAT ABOUT THE TECHNOLOGICAL 'PROOFS' OF RELATIVITY AND QUANTUM PHYSICS - GPS, LASERS, LEDS, CELL PHONES AND ALL SORTS OF ELECTRONICS ETC.?

It is claimed that the working of the GPS is the live proof of correctness of Special theory of Relativity as well the General theory of Relativity, the fact is, it's absolutely incorrect, on the contrary, I have already proved that the successful working of the GPS is the proof of the failure of Relativity which is already in public domain.

It is claimed that the working of lasers, LEDs, MRI scanners, transistors and therefore all integrated circuits, cell phones, computers and all electronics are based on Quantum Physics, so if Quantum Physics is wrong then how these devices are working fine? Conversely, it is claimed that, the very fact that these devices are working perfectly proves that Quantum Physics is correct. The answer is, Quantum Physics claims that the aim of Quantum Physics is not to know about the physical world, it doesn't exist at all, its aim is only what can be said about the microscopic quantum world, Quantum Physics is only interpretations of observations, Quantum Physics does not have only a single interpretation, as it is expected from any scientific theory, rather, there are many bizarre interpretations, every interpretation completely different from the other, all absolutely inconsistent with the Classical Physics. Why not it is possible that there may be completely new interpretations of the quantum world (NOT quantum mechanics) which is consistent with the Classical Physics at fundamental level and which is devoid of mysteries and magic that are present in all the present explanations? In fact, it has already been discovered, it only awaits to be revealed to the world, the details are given in my book – The Concepts, The Dawn of a New Physics – The Quassical Physics.

If one believes that the present state of Quantum Physics is correct then one should also believe his cell phone doesn't exist, like the moon doesn't exist, unless it is looked upon or it is called, and one should also be ready to believe that every call one receive in his cell phone is in an entirely new parallel universe, how easy it has become to create a whole new universe, just a missed call, or even an sms can do, thanks to the Quantum Physics. The fact is, these claims are made by the people who don't understand the postulates and their implications of the Quantum Physics, but the problem is – who understands the Quantum Physics? Remember the statement of Feynman, I suppose, nobody has contradicted Feynman so far. The answer lies in the fact that, technology is not bothered to understand if it is based on any theoretical principles or it is contradicting some theoretical principles, it is concerned only with the practical working model and it is not the job of technologist to understand the theory behind it also. It is a fact that no quantum physicist has any understanding that how the quantum world works, everyone accepts this, but not understanding any phenomenon by the theorist does not prevent its technological use, it doesn't stop the Physics to work. If we look closely, in history and even in present, a vast amount of technology is in use without the complete theoretical understanding behind them, for example, people have been using the water wheels to perform some king of work for ages, the water wheel can only work if there is a gravitational field, the waterwheel exploits the gravitational potential energy to extract some energy, so indirectly this is also a technological use of gravitation, people understood the properties of gravitation, what gravitation can do without any understanding how it works, this is sufficient to make the waterwheel and exploited the properties of gravitation for their use, they do not need what is the correct theory behind it. The field equations of the General Relativity is not important for the technological use of gravitation, even today, is there anybody on Earth who understands conceptually how the field equations of the General Relativity creates gravitation? There is nobody, because the field equations do not provide any conceptual understanding of gravitation, it only claims to provide mathematical knowledge and that is not physics. There exist people who claim that they understand the field equation of General Relativity and further claim that they understand how the equations completely explains the gravitation, there is no dearth of people who can visualize the 4-dimensional curved space-tine creating gravitation. Similarly the understanding of the Schrodinger's equation was not necessary to understand the properties of the elements and chemical substances, the whole chemistry was working on empirical knowledge, it was exploited as technology, the whole periodic table was discovered before the formulation of Schrodinger's equation. Think about the photon and electron, the backbone of the modern technology, the optical fibers for high speed high bandwidth communication and the whole electronics, all are based on the empirical properties of photons and electrons, Quantum Physics does not have any clear understanding what a photon is and what an electron is, the mystery of dual nature of photon as well as an electron are still there to haunt. These technological devices use the empirical knowledge which make them work and are completely uninfluenced and independent of which version of theoretical interpretation of Quantum Physics is correct, since they are based on empirical knowledge, they remain unaffected and will keep working even if all the interpretations of present quantum theory is proved completely wrong on theoretical grounds. Therefore, these claims are irrelevant; they are not related in any way with the scientific proof of correctness of the Quantum Physics.

25
THE ACCEPTED CONCEPTS OF THE FUNDAMENTAL THEORIES OF MODERN PHYSICS WHICH ARE NON-SCIENTIFIC

Following are the list of accepted concepts of the fundamental theories of Modern Physics which are non-scientific in the author's view, some of them are mentioned in this book also but each and every item is discussed in the corresponding books in details:

Special Relativity:
- The concept of time as an equivalent dimension of space
- 4-dimensional space-time
- First postulate of special relativity
- Second postulate of special relativity
- Lorentz's transformation
- Length contraction
- Time dilation
- Mass swelling
- Relative simultaneity

General theory of Relativity
- Equivalence principle
- 4-dimensional Curved space-time
- Curved space-time is Gravity
- Mass creates curvature of space-time
- Space-time guides a mass to move
- Field equation of general relativity

Book 1: Why A New Revolution in Physics is Essential and why it is Inevitable

    Gravitational time dilation
    Gravitational red shift
    Gravitational waves
    Black hole
    Frame dragging and geodetic precession

Quantum Mechanics
    Uncertainty principle
    The whole concept of wave function
    Probability amplitude
    Quantization of angular momentum
    Spin
    Quantum numbers
    Orbitals
    Hybridization
    Atomic and molecular bonding
    Self-interference
    Normalization
    Dirac Notations
    Entanglement
    Superposition

Standard Model and QFT
    The concept of strong force
    The concept of Weak force
    The families of particles
    The particles zoo
    Quantum fields
    Fundamental particle is mode of vibration
  … and more.

26

WHAT IS THE BASIS OF THE QUASSICAL PHYSICS? HOW ITS BASIS IS DIFFERENT FROM THE BASIS OF THE PRESENTLY ACCEPTED PHYSICS

The new physics – the Quassical Physics, is based on empirical evidences of the real physical world obtained through the experiments which have already been performed, some of these important experiments are given below, and some obvious empirical facts are also taken into account:
1. For Nature of light / electromagnetic wave:
    a. Propagation of light through Vacuum, from the Sun to the Earth the obvious fact.
    b. Michelson Morley experiment
    c. Arago's experiment
    d. Fizeau's experiment
    e. James Bradley's Aberration experiment
    f. Maxwell's derivation of speed of light
    g. Sagnac experiment
    h. Black body radiation characteristics
    i. Hertz's Experiment
    j. Compton experiment
    k. Photoelectric effect experiment
    l. Single slit experiment
    m. Young's double slit experiment
    n. Double refraction experiments
    o. Spontaneous Parametric conversion

Book 1: Why A New Revolution in Physics is Essential and why it is Inevitable

   p. Hydrogen emission spectra
   q. Newton's dispersion experiment
   s. Partial reflection and refraction of light rays.
   t. Light Polarization experiments
   u. Laser
   v. Femtosecond pulsed laser
   w. Laser cooling
   x. Olber's paradox
   y. Astronomical red shifts
   z. Red shift anomalies
   …and some more experiments and empirical facts
2. For Gravitation:
   a. Our own experience of weight
   b. Galileo's thought experiment about falling mass
   c. Newton's thought experiment about falling apple
   d. Cavendish's Torsion Pendulum
   e. The solar system
   f. The simple Pendulum
   g. Eotvos Experiment
   h. Light curve of a galaxy
   i. Newton's nightmare - Action at a distance
   p. Geostationary satellites
   q. Astronomical observations: stability of star clusters, galactic clusters
   r. Astronomical observation: structure of spiral galaxies.
   s. Astronomical observation: Hubble deep field
   …and some more experiments and empirical facts
3. Structure of Atom
   a. Thomson experiment
   b. Rutherford Experiment
   c. James Chadwick Experiment
   d. Hydrogen emission spectra
   e. Multi electron atom emission spectra
   f. Photoelectron Spectroscopy
   g. Stern Gerlach Experiment
   h. Davisson Germer experiment
   i. Zeeman effect experiment
   j. Stark Effect experiment
   k. Photo electric effect experiment
   l. Thermionic emission experiment

- m. Ionization energy experiments
- n. The modern Periodic Table, The biggest empirical fact of physical world.
- o. X-ray tube
- p. Modern High power gap electrode light bulb
- q. Atomic and molecular bonding,

...and some more experiments and empirical facts.

As mentioned above the basis of the Quassical Physics is entirely empirical, it is based on the primacy of physical understanding of a physical phenomenon, rather than on abstract mathematical equations, only very basic and minimal mathematics is used, while, the basis of the Modern Physics is NOT empirical, although it is claimed so, it is based on non-empirical, non-scientific personal views and primacy of mathematical equations, the importance of physical understanding of a physical phenomenon is absolutely rejected, "shut-up and calculate" summarizes this philosophy.

27
HOW THE QUASSICAL PHYSICS AFFECTS THE PARTICLE PHYSICS

Since, the Standard Model is mainly based on Quantum Mechanics, it is severely affected. As a side effect of the Quassical Physics, the whole particle physics will completely evaporate, no standard model, no three families of particles, no force carriers, no strong force, no weak force, no electroweak force, no zoo of particles, no quarks and no Higgs boson.

28
HOW THE QUASSICAL PHYSICS AFFECTS THE MATHEMATICS

This revolution is not limited to the fundamentals of Physics only; it will affect the foundation of the mathematics also because mathematics is heavily used in modern physics. Like physics, the philosophical problem of foundation of mathematics is also very less known and some of the mathematicians who are aware, they prefer to ignore it and even some of the mathematicians, I would like to refer David Hilbert, who clearly warned about the problems of foundation of mathematics but no physicist listened to it, on the contrary the physicists let the mathematicians drive the modern revolution of physics, this is one of the main reason why the epicycles came into existence and kept on developing for over a century. My 2 books are on mathematics, I claim, that these two books will change the perception of mathematics forever, and will also redefine the way mathematics should be used in physics to minimize the distraction of progress of physics and to curb the possibility of any new epicycles..

29
HOW THE QUASSICAL PHYSICS AFFECTS THE PHYSICS ESTABLISHMENT, THE MAINSTREAM PHYSICISTS AND THE PEER REVIEW PROCESS?

The claims made by the Quassical Physical about the modern physics even in this first book of the series, provide a very strong reason for introspection to the Physics Establishment, something has very seriously gone wrong and, nobody can dare to point out on the fear of losing the comfort to remain within the System, and going outside the System is a sure death.

The books of the series "A New Revolution in Physics" creates the required unbiased, critical and scientific environment to the mainstream physicists and the drivers of the peer review process to openly ask the questions they are supposed to with less fear, they should ask themselves within the system, are they really following true scientific principles for the progress of physics? Aren't they following the non-scientific, non-philosophical, arrogant culture the founders have set in the beginning of the 20th century? Could they see the crises in physics? Could they realize how giant the crises have become? Could they at least see and read the solution of the crises presented by the new Quassical Physics? Shouldn't they analyze the presented solution of the crises scientifically?

My claims can be rejected outright, but the issues I have raised in these books will not die down easily, because they contain the elements of truth, sooner or later, they will have to be accepted, not because they are my views but because they are completely based on true scientific principles, they are open to be challenged on scientific and philosophical grounds, these seem the only alternative which not only solves all the problems created by our modern fundamental theories but also address the other very basic problems of Classical Physics which were buried long back, the Quassical Physics solves a very wide spectrum of problems and paradoxes. It seems it will be very difficult to find the scientific alternative to this new Quassical Physics in near future, unless some very strange new characteristics of the Nature are discovered. This is the natural way of the progress of physics, Nature does not reveal everything at once, it's a gradual process based on the interplay between the Nature and the state of technology. Every new critical discovery may affect the previous theories but if true scientific principles are followed then the effect will be minimal or it is possible that there may be no effect at all, but if scientific principles are not followed in the development of the fundamental theories then they may bring a paradigm shift of knowledge claiming it as understanding and finally sooner or later they will be forced to suffer another paradigm shift of knowledge by the empirical evidences and true scientific efforts, in between stalling the progress of Physics and wasting huge resources and efforts. Unfortunately, the Physics Establishment / the System is going through this phase, and if taken scientifically and critically, the Quassical Physics has everything to come out of this situation and take the path the giants have shown.

.

30
THE NEXT BOOK TO BE PUBLISHED "THE COLLAPSE OF THE QUANTUM PHYSICS"

The next book of this series is "The Collapse of the Quantum Physics", which highlights non-scientific steps taken in its developments, some of which are already discussed in this book, it describes the complete failure of the Quantum physics on philosophical, theoretical, mathematical and experimental fronts which calls for the necessity of the new revolution in Physics.

31
CONCLUSIONS

A very brief history of the development of the fundamental theories of the Modern Physics has been presented in this book, and many non-scientific steps taken by the founders of the theories of the Modern Physics are highlighted in brief, they introduced a non-scientific philosophy in physics, some of their comments are quoted which also throw some light what they thought and how their personal views influenced the development of the fundamental theories, they preferred their non-scientific personal views over the scientific requirements, and deviated from the aim of the true physics – to understand the Nature, the real physical world, and took a path of mathematical dominance without any physical understanding and have finally killed the Classical Physics, as well as the true culture of Physical understanding by scientific methods. The knowledge management system followed these non-scientific culture and finally at present we have two great epicycles – the theory of Relativity and the theory of Quantum Mechanics.

Based on scientific analysis, the author claims that these two main theories are absolutely wrong, this book only provides a glimpse of it and the complete details are discussed in separate books, this creates a new revolution in Physics. The author also claims that the cure of these epicycles have been found which needs a new physics – the Quassical Physics, the blend of scientific ideas of the Classical Physics and Quantum Physics (not Quantum Mechanics) and a few new postulates, which solves almost all the problems of Relativity and Quantum Physics

Book 1: Why A New Revolution in Physics is Essential and why it is Inevitable

by providing alternative explanations, as well as many of the problems of the Classical Physics which were buried by the Modern Physics, this makes the new revolution inevitable.

With this book a new revolution in Physics has started, purely on scientific grounds, let us participate and clean the physics and restore the true sense of physics in Physics.

www.ingramcontent.com/pod-product-compliance
Lightning Source LLC
Chambersburg PA
CBHW030014190526
45157CB00016B/2704